Zhiming Chen

Modelling the plastic deformation of iron

**Schriftenreihe
des Instituts für Angewandte Materialien**

Band 15

Karlsruher Institut für Technologie (KIT)
Institut für Angewandte Materialien (IAM)

Eine Übersicht über alle bisher in dieser Schriftenreihe erschienenen Bände
finden Sie am Ende des Buches.

Modelling the plastic deformation of iron

by
Zhiming Chen

Dissertation, Karlsruher Institut für Technologie (KIT)
Fakultät für Maschinenbau, 2012
Tag der mündlichen Prüfung: 9. Juli 2012

Impressum

Karlsruher Institut für Technologie (KIT)
KIT Scientific Publishing
Straße am Forum 2
D-76131 Karlsruhe
www.ksp.kit.edu

KIT – Universität des Landes Baden-Württemberg und
nationales Forschungszentrum in der Helmholtz-Gemeinschaft

KIT Scientific Publishing 2013
Print on Demand

ISSN 2192-9963
ISBN 978-3-86644-968-8

Modelling the plastic deformation of iron

Zur Erlangung des akademischen Grades

Doktor der Ingenieurwissenschaften

Der Fakultät für Maschinenbau

Karlsruher Institut für Technologie (KIT)

genehmigte

Dissertation

von

M. Sc. Zhiming Chen

aus China

Tag der mündlichen Prüfung: 09 July 2012

Hauptreferent: Prof. Dr. rer. nat. Peter Gumbsch

Korreferent: Prof. Dr. -Ing. Erik Bitzek

Acknowledgements

It would not have been possible to write this doctoral thesis without the help and supports from all kind people around me, to only some of whom it is possible to give particular mention here.

First and foremost I offer my sincerest gratitude to my supervisor, Prof. Dr. rer. nat. Peter Gumbsch, who has supported me throughout this dissertation with his patience and encouragement. I attribute my level of Ph.D. degree to his excellent scientific guidance, never-ending encouragement and willingness to share his extensive knowledge and invaluable experiences. This work would not have been possible without his support. His limitless interests and enthusiasm in scientific research serve as a deep source of inspiration for me.

I would also like to express my greatest thanks to Dr. Matous Mrovec, for his patient guidance throughout my whole work, valuable discussions and countless help in developing my extensive research skills and shaping my scientific writing.

I would like to thank Prof. Vaclav Vitek from University of Pennsylvania for providing his valuable suggestions and deep scientific insight. I am also indebted to Dr. Roman Gröger from Academy of Sciences of Czech Republic for his comments and suggestions. Many thanks are due to Dr. Daniel Weygand, Prof. Dr. -Ing. Erik Bitzek and Kinshuk Srivastava for helpful discussions and valuable advices.

I am indebted to Prof. Ke Lu from Chinese Academy of Sciences and Prof. Zhaohui Jin from Shanghai Jiaotong University, who provide me the initial impulse of research and endless encouragement to pursue my Ph.D. degree in Germany.

The members of the Gumbsch group have contributed immensely to my professional time at Karlsruhe. The group has been a source of friendship as well as good advices and collaborations. I owe my greatest appreciation to Dr. Christoph Eberl, Rudolf Baumbusch, Dr. Jochen Senger, Melanie Syha, Dr. Matthias Weber, Dr. Sandfeld Stefan, Dr. Diana Courty, Dr. Katrin Schulz, Dr. Thomas Gnielka, Valentina Pavlova, Jia Lin and all other current and former group members. Special thanks are due to Dr. Dmitry Bachurin, for sharing most work hours in the same office with help and good humor that delight my stay in the past 4 years. Many thanks should be directed towards Mrs. Andrea Doer and Mrs. Daniela Leisinger for their help with many administrative issues and to Mrs. Yiyue Li for her technique support.

I also appreciate my Dissertation committee, Prof. Dr. rer. nat. Peter Gumbsch and Prof. Dr. -Ing. Erik Bitzek, for their time, interests, helpful comments and suggestions.

The funding from Karlsruhe Institute of Technology that made my Ph.D. work possible is greatly acknowledged.

My time at Karlsruhe was made enjoyable in large part due to all my friends here, with whom I am grateful for time spent together.

Finally, my greatest thanks belong to my parents, for their unconditional love and endless supports, which provide all my inspiration and are forever my driving force.

Abstract

The plastic deformation of body-centered cubic (bcc) iron at low temperatures is governed by the $a_0/2\langle111\rangle$ screw dislocations. Their non-planar core structure gives rise to a strong temperature dependence of the yield stress and overall plastic behavior that does not follow the Schmid law common to close-packed metals. In this work the properties of the screw dislocations in Fe is studied by means of static atomistic simulations using a state-of-the-art magnetic bond order potential (BOP). The core structures at equilibrium as well as under various external loadings are examined. Based on the atomistic studies an analytical yield criterion is formulated that captures correctly the non-Schmid plastic response of iron single crystal under general deformation. The yield criterion was used to identify operative slip systems for uniaxial loadings in tension and compression along all directions within the standard stereographic triangle. A good agreement between our theoretical predictions and experimental data demonstrates the robustness and reliability of such atomistically-based yield criterion. In order to develop a link between the atomistic modeling of the $a_0/2\langle111\rangle$ screw dislocations at 0 K and their thermally activated motion via nucleation and propagation of kinks at finite temperatures, a model Peierls potential is introduced which is able to reproduce all aspects of the dislocation glide resulting from the non-planar core structure. Using the transition state theory, the predicted temperature dependences of the yield stress as well as some characteristic features of the non-Schmid behavior such as the twinning-antitwinning and tension-compression asymmetries agree well with experimental observations. The results presented in the thesis therefore establish a consistent bottom-up model that provides an insight into the microscopic origins of the peculiar macroscopic plastic behavior of bcc iron at low temperatures. In addition, the results obtained in this work can be utilized directly in mesoscopic modeling approaches such as discrete dislocation dynamics.

Kurzfassung

Bei niedrigen Temperaturen wird die plastische Verformung von kubisch-raumzentriertem (krz) Eisen durch $a_0/2$<111> Schraubenversetzung kontrolliert. Ihre nicht-planare Kernstruktur führt zu einer großen Temperaturabhängigkeit der Fließspannung und das gesamte plastische Verhalten lässt sich nicht durch das für dichtgepackte Metalle geltende Schmidgesetz beschreiben. In dieser Arbeit werden die Eigenschaften von Schraubenversetzungen in Eisen mit Hilfe einer statischen Atomistiksimulation untersucht, das ein sich auf dem aktuellen Stand der Technik befindendes magnetisches Bindungspotential, das sogenannte "Bond-Order Potential" verwendet. Die Kernstruktur wird bei Gleichgewicht und unter verschiedenen externen Belastungszuständen untersucht. Basierend auf den atomistischen Untersuchungen wird ein analytisches Fließkriterium formuliert, welches das sogenannte "non-Schmid" Verhalten von einkristallinem Eisen bei allgemeiner Verformung wiedergibt. Das Fließkriterium wird verwendet, um die aktiven Gleitsysteme bei einachsiger Belastung im Zug und Druck entlang aller Richtungen des stereographischen Dreiecks herauszufinden. Eine gute Übereinstimmung zwischen unseren theoretischen Vorhersagen und den experimentellen Daten zeigt die Robustheit und die Zuverlässigkeit des Fließkriteriums basierend auf der atomistischen Simulation. Um eine Verbindung zwischen der atomischen Modellierung von $a_0/2$<111> Schraubenversetzung bei 0 K und ihrer thermisch aktivierter Bewegung bei endlichen Temperaturen durch die Nukleation und Ausbreitung von kink-paaren zu knüpfen, wird ein Peierls potential Modell eingeführt, welches alle Aspekte des Versetzungsgleitens resultierend aus der nicht-planar Kernstruktur reproduzieren kann. Unter Verwendung der "Transition State Theory" stimmen sowohl die vorhergesagten Temperaturenabhängigkeiten als auch die bestimmten charakteristischen Eigenschaften des "non-Schmid"-Verhaltens, wie die Zwillings-Antizwillings- sowie die Zugdruckasymmetrie sehr gut mit experimentellen Beobachtungen überein. Die in diese Arbeit präsentierten Ergebnisse bauen ein konsistentes "Bottom-up" Modell auf, das einen Einblick in den mikroskopischen Ursprung des eigenartigen makroskopischen plastischen Verhaltens von krz Eisen bei niedrigen Temperaturen liefert. Zusätzlich können die in dieser Arbeit erzielten Ergebnisse in mesoskopische Modellierungsansätze wie die diskrete Versetzungsdynamik direkt übertragen werden.

Contents

Abbreviations

AM	Ackland and Mendelev potential
bcc	body-centered cubic
BOP	bond order potential
CRSS	critical resolved shear stress
DDD	discrete dislocation dynamics
DFT	density functional theory
DOS	density of states
EAM	embedded atom method
EI	elastic interaction
Exp	experiment
fcc	face-centered cubic
FIRE	fast inertial relaxation engine
FS	Finnis-Sinclair potential
hcp	hexagonal close packed
HRTEM	high-resolution transmission electronic microscopy
LP	large positive
LT	line tension
MD	molecular dynamic
MEP	minimum energy path
MRSSP	maximum resolved shear stress plane
NEB	nudged elastic band
PES	potential energy surface
RSS	resolved shear stress
SD	strength differential factor
SN	small nagative
SP	small positive
TB	tight-binding
TEM	transmission electron microscopy
TST	transition state theory

List of Symbols

a_0	lattice constant
a	kink height
a	lattice parameter of *m*-function
a_i	parameters of the analytical yield criterion
$\boldsymbol{b}$	Burgers vector
C_i	fitting parameters of the Peierls potential
C_{ij}	elastic constants
$dd\sigma$	bond integral of *sigma* molecular orbital
$dd\pi$	bond integral of *pi* molecular orbital
$dd\delta$	bond integral of *delta* molecular orbital
E	line tension of dislocation
E_b	dislocation energy
$E_{binding}$	bingding energy
E_{bond}	bond energy
E_{coh}	cohesive energy
E_{rep}	electrostatic and overlap repulsive energy
E_{mag}	magnetic energy
$\mathbf{F}$	force on NEB images
$\hat{f}$	flat top operator
$H(\sigma)$	stress-dependent activation enthalpy
H_k	energy of isolated kink
H_{kp}	activation enthalpy
k	spring constant
k	Peierls potential parameter
k_B	Boltzmann constant
$K_\sigma(\chi)$	σ dependent Peierls potential parameter
$K_\tau(\chi)$	τ dependent Peierls potential parameter
$\mathbf{l}_\alpha$	dislocation line direction
$\mathbf{n}_\alpha$	slip plane

$\mathbf{n}_\alpha^1$	reference slip plane
$\mathbf{m}_\alpha$	slip direction
$\mathbf{R}$	nudeged elastic band coordinate
r_{ij}	atom distance
T	temperature
T_k	knee temperature
u	flat top factor
V	Peierls potential
V_0	maximum height of the Peierls potential
V_σ	σ dependent Peierls potential
V_τ	τ dependent Peierls potential
$\mathbf{v}_i$	i component of dislocation velocity
$\mathbf{v}$	total velocity
W	work
Δz	kink-pair width

α	slip system index
χ	angle between MRSSP and $(\bar{1}01)$ plane
$\dot{\gamma}$	strain rate
η_α	loading path
λ	angle between loading direction and Burgers vector
μ	shear modulus
θ	Peierls potential parameter
σ	shear stress component parallel to the Burgers vector
$\bar{\sigma}$	athermal stress
σ^*	applied stress
σ_p	Peierls stress
σ_t	critical stress for tension
σ_c	critical stress for compression
Σ	stress tensor in MRSSP coordination
τ	shear stress component perpendicular to the Burgers vector
τ_{cr}^*	critical shear stress for yield criterion
$\hat{\tau}_i$	tangent vector between NEB images

ξ dislocation coordinate

ξ_0 initial dislocation position

ξ_c critical dislocation position

ψ angle between slip plane and $(\bar{1}01)$ plane

1 Introduction

1.1 Historical background and experimental overview

Starting from the 1920s, the earliest systematic studies of the mechanical properties of single crystals concentrated on the plasticity of the hexagonal close packed (hcp) and face-centered cubic (fcc) metals. The basic findings on the plasticity of these metals were (i) an essentially athermal nature of the deformation by crystallographical slip, and (ii) a universal description for the onset of slip known as the Schmid's law [1] (for review see [2, 3]). According to the Schmid law, the yield on the slip plane for a particular material occurs at a constant projected shear stress called the critical resolved shear stress (CRSS). The CRSS value depends neither on the slip system nor on the sense of slip. Additionally, the Schmid law assumes that the resolved shear stress on the activated slip system in the direction of slip (this stress is usually called the Schmid stress) is the only stress component triggering the plastic flow; the other components of the stress tensor do not affect the plastic deformation.

However, almost simultaneous investigations on α-iron [4] and β-brass [5] conducted by G.I. Taylor and his co-workers indicated that the slip behaviour in materials with the body-centered cubic (bcc) structures was completely different from that of the close-packed fcc and hcp metals, indicating that the Schmid law is not universally applicable. In the course of time, owing to the development of modern techniques for growing crystals and purification [6], extensive investigations have been performed on a broad variety metals and alloys with bcc structure including α-iron and iron-silicon alloys [7-21], the refractory metals of Groups VB and VIB [22-35], and the alkali metals [36-41]. All these experimental studies have unequivocally shown that all bcc materials exhibit certain common features in their deformation behavior, which distinguish the whole group from the fcc and hcp metals and alloys. These general features include: 1) a rapid increase of the yield and flow stresses with decreasing temperature and increasing strain rate, 2) a strong dependence of the CRSS on the orientation of the loading and overall breakdown of the Schmid law in single crystals of bcc metals [42-47].

Two distinct intrinsic non-Schmid effects have been observed in bcc metals [44]. The first one is the variation of the CRSS with the sense of shear called the twinning/anti-twinning asymmetry. The second one is the tension–compression asymmetry [48], where the critical stresses for tension and compression are different for the same loading orientation. It is now generally accepted that the origin of both these non-Schmid effects is related to a non-planar core structure of the $a_0/2<111>$ screw dislocations in the bcc lattice. The fact that the mechanical behaviour and flow stress of bcc metals below the so-called knee temperature T_k [49] is governed by the motion of the $a_0/2<111>$ screw dislocations was first proposed by Hirsch in 1960. These dislocations were expected to possess a high lattice friction as a consequence of the three-fold symmetry of the dislocation core structure in the bcc lattice [50]. This concept was supported later by transmission electron microscopy (TEM) observations in which long screw dislocation segments were observed in samples deformed at low temperatures (for example, see [51]). After decades, being confirmed by many experimental and theoretical studies, it is now well accepted that the strong temperature, strain-rate, and orientational dependence of the flow stress indeed results from intrinsic properties of the $a_0/2<111>$ screw dislocations at the atomic scale (for reviews see Refs. [52-56]). In order to fully understand the macroscopic mechanical behaviour of bcc iron, it is therefore necessary to analyze and describe the properties and the behaviour of the screw dislocations at the nanoscale, both with and without externally applied loadings. This investigation is one of the main topics of this thesis presented in Chapter 3.

A number of experimental studies examining fundamental characteristics of slip deformation in iron at low temperatures had been done in the past by, e.g. Allen et al. [57], Basinski and Christian [58], Conrad and Scheock [59], and Refs. [18, 60-67]. However, since the plastic deformation by dislocation slip is at very low temperatures competing with deformation by twinning or even with cleavage fracture, it took rather long time before the fundamental aspects of dislocation plasticity, such as the yield stress for slip or the operating slip systems, have been examined in the whole temperature range. Only in 1981, Aono et al. carried out a detailed systematic study of deformation of Fe single crystals down to liquid He temperatures [68]. The authors showed that the deformation mode depends on the specimen size and they succeeded to plastically deform specimens of smaller sizes at temperatures of 4.2 K and to investigate various fundamental characteristics of slip deformation. Similar to most other bcc metals, the twinning-antitwinning effect was also observed in Fe with the twinning-antitwinning ratio ranging between 1.12 and 1.22 for different loading directions. The variation of the yield stress as a function of temperature and orientation showed also characteristic features of bcc deformation behavior being divided into three temperature ranges: $T < 100$ K, 100 K $< T <$ 250 K, and 250 K $< T < 340$ K [69, 70]. Above 340 K lies the athermal region in which the value of yield stress was ~15 MPa independent of loading orientation. The regime between 250 K and 340 K is assigned to the fully developed kink-pairs governed by the elastic-interaction (EI) approximation [49]. Below 100 K it is governed by the formation of kink-pairs in a manner of bow-out on the primary {110} slip plane according to the line-tension (LT) approximation [71]. Regarding the regime between 100 K and 250 K, there is a discrepancy in the basic slip mechanism. While Brunner and Diehl argued that in this region the screw dislocations glide on {110} planes alternatively [72], in Seeger's explanation the kink-pair formation is assumed to be on the {112} planes [49].

The first attempts to determine the elementary slip mechanisms of the screw dislocation go back to studies of Taylor and Elam [4], who introduced the pencil glide mechanism where the slip was assumed to occur in the <111> crystallographic direction but the mean plane of slip was the one with the maximal projected shear stress. This plane might be a crystallographic but also a non-crystallographic plane. After this pioneering research, there have been several contradicting statements regarding the active slip planes in bcc metals [73]. Gough [74] and Barrett et al [75] stated that the slip takes place on the {110}, {112}, and {123} families of crystallographic planes. Other studies claimed that only the {110} and {112} slip planes are activated at ambient temperatures, whereby the {123} planes need a higher temperature for activation. More recent studies suggest that the elementary slip at the microscopic level takes place exclusively on the {110} planes, and the apparent slip on both the {112} and

{123} planes is actually composed of multiple elementary slip steps on two non-parallel {110} planes [76]. A systematic observation of the slip planes in single crystal iron was also presented by Aono et al. [68]. According to their results, the deformation below 200 K is clearly governed by the screw dislocations whose slip plane is exclusively the $(\bar{1}01)$ plane at liquid He temperature for any loading orientation with straight slip lines parallel to each other. However, as temperature being increased, the macroscopically observed slip plane approaches the maximum shear stress plane.

Another interesting feature observed in experiments was the phenomenon of anomalous slip [77, 78]. The anomalous slip occurs in bcc crystals at low and moderate plastic strains when the deformation proceeds on a set of {110} planes on which the resolved shear stress is substantially lower than that on the primary, i.e. with the highest Schmid factor, {110} slip plane. All these experimentally observed phenomena can not be fully understood without knowledge of microscopic processes associated with the glide of the screw dislocations. In order to establish a link between the macroscopic mechanical properties and the dislocation core structure, our first task is to determine the elementary slip behavior of the $a_0/2<111>$ screw dislocation in iron at the atomic scale.

Unfortunately, direct observations of the atomic core structures of the $a_0/2<111>$ screw dislocation in bcc metals are difficult and only very few attempts have been made so far [79, 80]. This is because the atoms around a screw dislocation are displaced primarily along the dislocation line direction while their displacements perpendicular to the dislocation line, which can be detected by the modern high-resolution transmission electronic microscopy (HRTEM), are usually very small (their magnitude is given by the elastic anisotropy of the material). Thus, the understanding of the screw dislocation behaviour at the atomic level relies ultimately on the modelling and simulation techniques.

1.2   Modeling and simulations of screw dislocations in bcc metals

1.2.1   Intrinsic properties at 0 K

First computer-based atomistic modelling studies of the structure and energetic of the $a_0/2\langle111\rangle$ screw dislocations were carried out in the 1970s by Duesbery [26], Vitek et al. [81], and Basinski et al. [82] using simple pair potentials. Already these pioneering calculations revealed the non-planar core of the screw dislocation and confirmed thus the initial assumptions about the limited mobility of these line defects. As indicated by Neumann's principle [83], the main factor controlling the core spreading is the symmetry of the crystal structure. The most important symmetry consideration relevant to the $a_0/2\langle111\rangle$ screw dislocation in a bcc crystal is that $\langle111\rangle$ is the direction of a threefold screw axis. The first computer simulations showed that the screw dislocation core indeed possessed the three-fold symmetry and extended along three {110} glide planes containing the $\langle111\rangle$ slip direction. Two energetically degenerate conjugate configurations were found that were related to each other by a rotation of $\pi/3$ in terms of the $\langle111\rangle$ diad [Fig. 1-1(a)]. These core structures were later termed as *degenerate.* In the following years, the degenerate core structure of the $a_0/2\langle111\rangle$ screw dislocation was found in many bcc metals using more advanced many-body interatomic potentials such as the Finnis-Sinclair (FS) potential or potentials based on the embedded atom method (EAM) [54, 84-87]. However, it was found that there exists also another variant of the screw core termed as *non-degenerate*, which is characterized by the <110> diad symmetry operation. This non-degenerate core structure had already been found by some central-force potentials [88-92], and it has been primarily identified as the ground-state structure in recent calculations employing accurate first-principles methods based on the density functional theory (DFT) [93-97], closely related tight-binding (TB) models, and TB-based bond-order potentials (BOP) [98-100]. Based on the high credibility of these recent calculations, it is now thought that the non-degenerate core structure is indeed the equilibrium structure of the $a_0/2$<111> screw dislocation for most bcc metals [Fig. 1-1(b)].

The determination of the correct ground-state dislocation core structure is only the first step. In order to build a link to macroscopic plasticity, it is necessary to examine how the dislocation responds to externally applied loads (see, for example, [44, 101]). The purpose is to identify the components of the stress tensor that influence the motion of an individual screw dislocation and subsequently to quantify their effects on the magnitude of the Peierls stress. For fcc metals according to the Schmid law, the dislocation starts to move when the resolved shear stress (RSS) in the slip plane parallel to the direction of the Burgers vector reaches a critical value, i.e., CRSS (critical resolved shear stress). The Peach-Köhler force is in this case the only contribution driving the dislocation forward. The sense of the shearing and the components of the stress tensor other than the shear stress parallel to the slip direction in the slip plane have no effects on the dislocation motion [52]. In contrast to the fcc metals, it was proved by both experimental and atomistic studies [89, 101, 102] that the Schmid

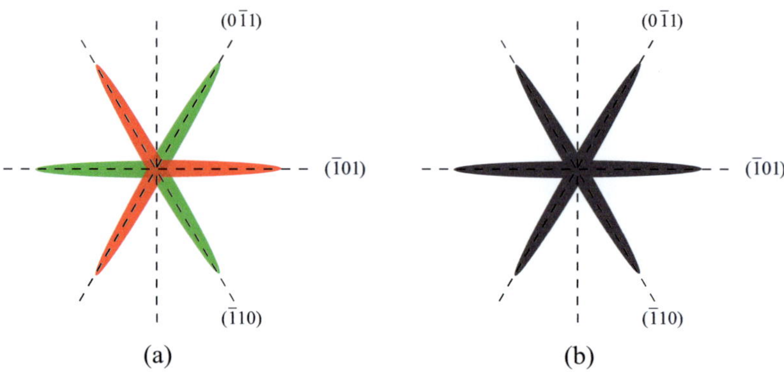

Figure 1-1. The *degenerate* (a) and *non-degenerate* (b) core structures of bcc lattice.

law is not valid in bcc metals and the glide of the screw dislocation is also significantly affected by stress components other than those parallel to the slip direction. Direct manifestations of these effects in bcc metals are the experimentally observed twinning-antitwinning and tension-compression asymmetries [44]. These macroscopic phenomena are already clearly visible at the atomic level. For example, in [101] a set of pure shear stresses parallel to the slip direction in different maximum resolved shear stress planes (MRSSP) were applied to the $a_0/2<111>$ screw dislocation to verify the twinning-antitwinning effect. In addition, uniaxial loadings in tension and compression with corresponding MRSSP were performed to examine the tension-compression asymmetry. The purpose of these calculations was also to reveal that the shear stress parallel to the slip direction, which exerts Peach-Köhler force to drive the screw dislocation, is not the only stress component controlling the motion of the dislocation. Instead, the shear stresses perpendicular to the slip direction, which do not drive the screw dislocation directly, can change the core structure, and consequently affect the CRSS of the screw dislocation.

Apart from the intrinsic properties of a single $a_0/2<111>$ screw dislocation that determine the onset of the plastic deformation in single crystals, continuum yield criterions are highly desirable for macroscopic engineering calculations. With such yield criterions the microscopic behaviour can be represented using a small number of fundamental parameters. The early framework of the continuum description for single crystal plasticity was developed by Hill [103] and Rice [104]. These theories are commonly based on the Schmid law for close-packed metals. These first models have been later extended to include the non-Schmid effects but only in a limited manner [105, 106]. More systematic work on the non-Schmid description has been developed in the early 1990's by Qin and Bassani for Ni_3Al [107, 108]. In their model, the critical resolved shear stress in the primary slip system is a function of both the orientation of the loading axis and the sense of shear. A simple form of an effective yield criterion was constructed in which the yield stress is written as a linear combination of the Schmid stress and other non-Schmid stresses. Recently, the analytical yield criterion proposed in [107, 108] was further developed by Gröger et al. [109] to determine the commencement of the motion of the $a_0/2<111>$ screw dislocations in Mo and W under general external loading. This yield criterion was shown to reproduce closely the atomistic results including the non-Schmid effects such as, the twinning-antitwinning and tension-compression asymmetries, and therefore to be able to determine reliably the slip behaviour of a single crystal under any loading orientation.

1.2.2　Finite temperature behavior

Most of the atomistic studies mentioned above are static calculations that provide information about an ideal, infinitely long and straight $a_0/2<111>$

screw dislocation at 0 K. Owing to the non-planar core structure, the lattice resistance is very large indicating a high Peierls barrier between two neighbouring sites that the dislocation has to overcome [52, 110-113]. The corresponding critical resolved shear stress required to surmount this barrier at 0 K is called the Peierls stress.

As mentioned above, it is observed experimentally that the yield stress for the whole group of bcc metals including Fe is not constant but it strongly depends on temperature. There exist two main regions of the yield stress in terms of temperature [49]: at temperatures $T > T_k$, above the so-called critical temperature T_k, the yield stress is only weakly dependent on temperature and can be approximated by a constant athermal stress $\bar{\sigma}$. For $T < T_k$ there is a rapid increase of the yield stress, σ^*, with decreasing temperature.

Seeger [49] suggested that the flow stress of metals with bcc structure is determined by the thermally activated formation of kink pairs on the $a_0/2<111>$ screw dislocations and their subsequent migration along the dislocation line. The model of the formation and propagation of kink-pairs was originally developed for fcc metals where the dislocation glides on a well-defined slip plane and the Peierls barrier is a periodic function of the dislocation position [114-117]. The application of this concept to bcc metals enabled to account for the temperature dependence of the flow stress, however, the model parameters need to be determined by empirical fitting to experimental data. Such a study was conducted by Brunner and Diehl [72, 118, 119], in which the high purity α-iron single crystals were measured via stress-relaxation measurements by successive tensile deformation steps at different temperatures from 4K to 380K [69, 70]. Several characteristic kink and dislocation parameters were evaluated quantitatively from the experiments providing a rather good agreement between the experimental results and the theoretical predictions for the dependence of the flow stress on temperature.

In Seeger's model, the screw dislocation is expected to surpass the Peierls barrier with the aid of thermal activation via nucleation of kink-pairs, which subsequently migrate relatively easier along the dislocation line [52, 120, 121]. Since a part of the energy required to activate the dislocation is supplied by thermal fluctuations, the corresponding CRSS required to trigger the motion of the screw dislocation in this way is thus smaller than the Peierls stress at 0K for the ideal straight desolation. Following the transition state theory of thermally activated processes [122-124], a simple Arrhenius equation representing the strain rate

$$\dot{\gamma} = \dot{\gamma_0} \exp[-\frac{H(\sigma)}{k_B T}] \tag{1-1}$$

can be applied for the description of the process. The pre-exponential factor $\dot{\gamma_0}$ depends on the details of the mechanism of kink-pair formation but can be considered to a good approximation as constant. $H(\sigma)$ is the activation enthalpy, which is a function of the applied stress tensor σ, k_B is the Boltzmann constant and T is the absolute temperature. Eq. 1-1 therefore describes the dependence of the flow stress on temperature, where the crucial quantity which governs the process is the stress dependent activation enthalpy $H(\sigma)$.

(1) High-temperature/low-stress regime

At high temperatures close to T_k, the critical yield stress is significantly lower than the Peierls stress at 0 K and approaches the athermal yield

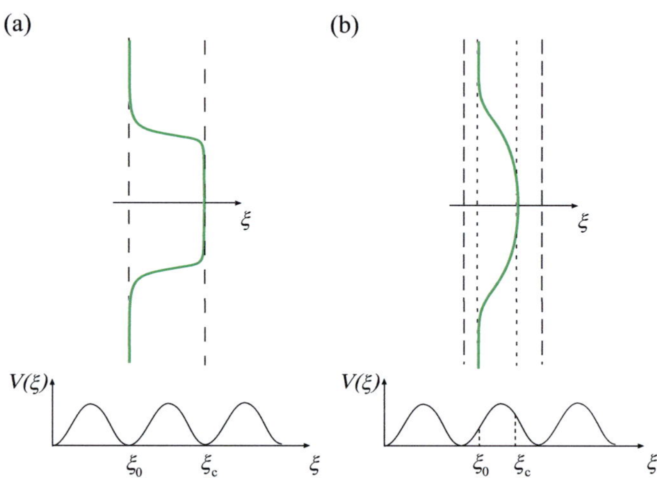

Figure 1-2. The saddle-point configurations for the nucleation of a pair of kinks on the $a_0/2<111>$ screw dislocation: (a) a pair of well-developed kinks at low stresses and high temperatures; (b) a bow-out at high stresses and low temperatures. ξ is the coordinate along the activation path and $V(\xi)$ is the Peierls barrier along this path.

stress. In this region, the thermal fluctuations are so large that fully developed kink pairs can form on the dislocations [See Fig. 1-2(a)].

The formed kinks interact elastically with each other via the attractive Eshelby potential which can be expressed as:

$$-\mu a^2 b^2 / 8\pi\Delta z \tag{1-2}$$

in the framework of isotropic elasticity [52]. In Eq. 1-2, μ is the shear modulus; a is the kink height corresponding to the distance between two neighbouring Peierls valleys; b is the magnitude of the Burgers vector, and Δz is the kink-pair width. During the formation of a kink-pair, the work done by the applied stress, σ^*, which is the shear stress projected on the glide plane and parallel to the slip direction, is $\sigma^* ab\Delta z$. The enthalpy associated with this configuration is then given as

$$2H_{\mathrm{k}} - \mu a^2 b^2 / 8\pi\Delta z - \sigma^* ab\Delta z \qquad (1\text{-}3)$$

where $2H_{\mathrm{k}}$ is the energy of two isolated kinks at zero stress, which can be determined from either simulations [125, 126] or experiments [118].

The competition between the attractive interaction between the kinks and the repulsive interaction produced by the loading, σ^*, during the nucleation depends on the separation distance Δz. Kink pair separated by less than the critical distance will attract each other and annihilate; more distantly separated kink pairs will continue to spread apart under the action of the applied stress.

The critical separation of kinks for which the enthalpy of Eq. 1-3 reaches the maximum defines the saddle-point configuration. The corresponding activation enthalpy is:

$$H_{\mathrm{kp}} = 2H_{\mathrm{k}} - (ab)^{3/2} \sqrt{\frac{\mu\sigma^*}{2\pi}} \qquad (1\text{-}4)$$

One should note that the shear stress σ^* is the only component of the applied stress tensor that drives the dislocation forward. The effects of other stress components, e.g., the shear stresses perpendicular to the slip direction, which may affect the Peierls potential, are not considered in this model.

Therefore the thermally activated motion of the screw dislocation in the low-stresses/high-temperature regime is therefore assumed not to be dependent on the height and the shape of the Peierls potential.

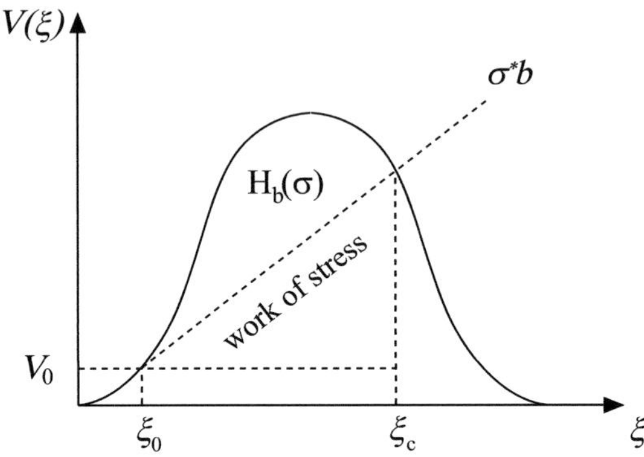

Figure 1-3. The activation energy H_b. $\sigma^* b$ is the Peach–Köhler force.

(2) Low-temperature/high-stress regime

At low temperatures, the thermal fluctuations are too small to be able to surmount the Peierls barrier. It is therefore necessary to lift the dislocation from the bottom of the Peierls valley with the aid of the applied stress. In the line tension model, the dislocation shift by the applied stress to position ξ_0 is determined from the condition [Fig. 1-2(b)]

$$\sigma^* b = \frac{\mathrm{d}V(\xi)}{\mathrm{d}\xi}\bigg|_{\xi=\xi_0} \tag{1-5}$$

At this position, the force pushing the dislocation back to the bottom of the Peierls valley is equal to the Peach–Köhler force $\sigma^* b$ due to the applied stress. The energy associated with the dislocation bow-out in a given Peierls potential is then given as [116]:

$$E_b = \int_{-\infty}^{+\infty} \{[V(\xi) + E]\sqrt{1 + \xi'^2} - [V(\xi_0) + E]\}\mathrm{d}z \tag{1-6}$$

where E is the line tension of the straight dislocation and the bowed segment is described by a function $\xi[x(z), y(z)]$, where the coordinate z lies along the slip direction. The work done by the bow-out is

$$W = \int_{-\infty}^{+\infty} \sigma^* b(\xi - \xi_0)\mathrm{d}z \qquad (1\text{-}7)$$

and the enthalpy associated with this process is then the difference $E_\mathrm{b} - W$. In most cases, the bow-out returns back to the original straight position. However if the saddle point is reached, corresponding to a critical value of $\xi = \xi_\mathrm{c}$, the bow-out continues to extend as a pair of fully developed kinks. Following Dorn and Rajnak [116], this leads to the following condition determining ξ_c:

$$V(\xi_\mathrm{c}) = \sigma^* b(\xi - \xi_\mathrm{c}) + V(\xi_0) \qquad (1\text{-}8)$$

The activation enthalpy of this configuration is finally given as

$$H_\mathrm{b} = 2\int_{\xi_0}^{\xi_\mathrm{c}} \sqrt{[V(\xi) + E]^2 - [\sigma^* b(\xi - \xi_0) + V(\xi_0) + E]^2}\,\mathrm{d}\xi \qquad (1\text{-}9)$$

where the Peierls barrier $V(\xi)$ needs to be determined either theoretically or from experimental data. The activation enthalpy H_b from Eq. 1-9 is schematically illustrated in Fig. 1-3 as an area bounded by the Peierls barrier and the line with the slope $\sigma^* b$. ξ_0 is the initial position of the dislocation under external loading and ξ_c can then be evaluated numerically according to Eq. 1-8.

(3) Determination of the activation enthalpy

A possible way to determine the activation enthalpy and its dependence on stress using atomistic simulations is to investigate the transition path between two neighbouring stable positions of the screw dislocation using the Nudged Elastic Band (NEB) method [127-130]. The NEB method is a powerful method for identifying the saddle-point configurations and can be applied even for complex activation processes. The procedure was adopted for degenerate cores in [131-134] and kink-pairs were observed in the commencement of the dislocation motion. The activation enthalpy for this process was then determined as a function of the pure shear stress applied along the $(\bar{1}01)$ glide plane.

However, as noted above, the Peierls barrier, and therefore also the activation enthalpy, can be influenced by all components of the applied stress tensor. This is especially true in the case of the $a_0/2{<}111{>}$ screw dislocations in bcc metals, whose core structure may be altered strongly by the non-Schmid stresses [101]. In order to take into account all contributions from the shear stress parallel as well as perpendicular to the slip direction, one would have to investigate the transition process for all possible combinations of the stress components. Such a procedure is obviously too computationally demanding and cannot be applied for the determination of the dependence of the activation enthalpy on the stress tensor.

An alternative approach for determining the temperature dependence of the yield stress is to study the motion of the $a_0/2{<}111{>}$ screw dislocations at finite temperatures directly by molecular dynamic (MD) simulations [91, 92, 133, 135, 136]. In these calculations the simulation block containing a single dislocation is loaded at a given temperature until the dislocations starts to move. The performed simulations show that the glide mechanism is indeed the nucleation and propagation of kink pairs on {110} planes, and the computed values of the flow stress decrease with increasing temperature. However, due to time and size limitations of MD simulations these calculations have to be performed at extremely high strain rates (typically 10^5 to 10^9 s^{-1} compared to 10^{-4} to 10^{-2} s^{-1} in experiments). Thus, MD simulations cannot be used for a systematic study of dislocation behaviour under realistic conditions.

Instead of the direct atomistic studies, Edagawa et al. [137] proposed in 1997 an analytical description of the Peierls potential, which is a periodical function of the position of the dislocation with three-fold rotating symmetry. Since this potential correctly satisfies the periodicity of the bcc lattice, it can be applied for analysis of the screw dislocation motion and the kink pair formation on any of the three adjacent {110} planes. For instance, the saddle-point configuration of a critical kink pair in three-dimensional space and the associated activation energy as a function of the shear stress parallel to the Burgers vector can be evaluated using the line-tension model of a dislocation [121, 122]. Recently, Gröger and his colleagues extended this model [138] and constructed the Peierls potential for bcc metals Mo and W based on results of atomistic simulations. The main advantage of this atomistically-based Peierls potential is that it correctly reflects features resulting from the non-planarity of the dislocation cores and its stress-induced transformations, e.g. the dependence of the Peierls stress on the MRSSP orientation and shear stresses perpendicular to the Burgers vector.

1.3   Objectives of this work

The main objective of this thesis is to develop a theoretical description of the low temperature plastic deformation governed primarily by the $a_0/2<111>$ screw dislocations in iron. To achieve this goal, the work starts with investigation of the basic properties of the straight $a_0/2[111]$ screw dislocation that are presented in Chapter 3. It focuses first on the effect of the pure shear stress parallel to the Burgers vector, to demonstrate the dependence of CRSS on the sense and orientation of the shearing and to reveal the so-called twinning-antitwinning asymmetry observed in experiments [44, 54]. Then the CRSS for the screw dislocation under uniaxial loadings in tension and compression is determined, in order to verify the experimentally observed tension-compression asymmetry. These calculations also reveal whether the shear stress parallel to the slip direction, which exerts the Peach-Köhler force to drive the screw dislocation, is the only stress component controlling the motion of the dislocation. As in [89, 102], the current work will prove that the non-Schmid stresses, i.e., the shear stresses perpendicular to the Burgers vector, also affect the dislocation motion in iron. These results will enable us to explain the anomalous slip observed in experiments, and to analyse in detail how the CRSS is influenced by the changes of the non-planar core structure of the $a_0/2<111>$ screw dislocation.

By applying reduced stress tensors with stress components only parallel and perpendicular to the slip direction, the critical resolved shear stress as a function of the orientation of the MRSSP as well as of the magnitude of the shear stress perpendicular to the slip direction is determined. These results can be used to determine the macroscopic yielding of single crystals containing $a_0/2<111>$ screw dislocations with all possible Burgers vectors. An analytical yield criterion will be formulated to determine the commencement of the motion of the $a_0/2<111>$ screw dislocations under external loadings at 0 K [109]. This criterion can predict the slip behaviour of Fe single crystal under any loading orientation.

In the last section of Chapter 3, a Peierls potential will be developed that captures all features resulting from the non-planarity of the screw dislocation core and its stress-induced transformations. Since the constructed Peierls potential is based on the results of atomistic simulations, it closely reproduces the dependence of the Peierls stress on the MRSSP orientation and on the shear stresses perpendicular to the Burgers vector. The thermally activated dislocation motion via formation of kink-pairs can then be treated using the line tension model at low temperatures and the elastic interaction model at high temperatures [121, 122].

Since direct atomistic simulations of dislocations by MD techniques are limited by small length and time scales, the understanding of the deformation behaviour at finite temperatures requires an employment of phenomenological models that describes only the key properties of dislocations, e.g., the dislocation mobility in terms of the activation enthalpy and the loading stress, instead of covering all atomistic details. In recent years, the newly developed Discrete Dislocation Dynamics (DDD) models provided a mesoscopic description of dislocation ensembles based on the single dislocation mobility. The dislocation mobility laws in most existing DDD models is based on Kocks law [120, 139], which describes the activation enthalpy as a function of the resolved shear stress by fitting the parameters to experiments. Such a mobility law cannot reflect the non-Schmid effects, e.g., the twinning-antitwinning and tension-compression asymmetries originating from non-planar cores of the screw dislocations in bcc metals. Thus, one of the ultimate goals of our work is to establish a bottom-up modelling approach which will enable transparent and well defined transfer of the achieved information on dislocation properties from the microscopic through the mesoscopic to the macroscopic level. This multiscale framework is elaborated in Chapter 4, where it starts from the fundamental dislocation properties at the atomic level and build up a link between the atomic-level modelling of the glide of $a_0/2<111>$ screw dislocations at 0 K and the mesoscopic modelling of the thermally activated motion of screw dislocations via nucleation of kink-pairs at finite temperatures. The approach developed here provides a consistent multiscale picture about the low temperature plastic deformation of bcc iron. The results obtained in our work can be utilized directly as input data in higher level modelling schemes such as DDD.

2 Methods

2.1 Bond order potential

One of the most critical aspects of all atomistic modelling and simulations is their dependence on the description of the interatomic interactions. This is particularly important near the defects such as vacancies and dislocations. Methods for evaluation of interatomic forces can be divided roughly into three classes. The first class employing a full quantum mechanical treatment of the electronic structure, for instance within DFT, provides the description of chemical bonding most reliably and have been employed in many investigations of the physical and mechanical properties of materials (for reviews, see e.g., [55, 140-142]). However, the first-principles calculations are limited to small block sizes and restricted by the use of periodic boundary conditions. Studies of large and complex systems, e.g., those with dislocations, typically require approximations and significant simplifications when describing the interatomic interactions. Such studies are nowadays mostly carried out using a second class of methods, namely empirical potentials. These methods are very computational efficient but often sacrifice reliability and transferability. For instance, many-body FS [143] or EAM [144, 145] potentials are able to describe well simple and noble metals, in which the bonding is almost nearly free-electron-like, but not transition metals such as iron, where the bonding between atoms is mediated by the *d*-electrons and further complicated by magnetic effects. In the present work, a third class of methods is therefore employ. This third class presents a compromise between the former two classes in terms of reliability and speed, and is hence best suited for studying the properties of the $a_0/2<111>$ screw dislocations in iron. The model used in this work is a recently developed magnetic bond-order potential (BOP) for Fe [99, 100], which is based on the tight-binding approximation to the electronic structure and therefore it is able to describe correctly the angular character of bonding in Fe. Despite its quantum mechanical origin, BOP is also sufficiently computationally efficient for the modelling of extended defects, and due to its real space formalism it is not limited by the periodic boundary conditions.

Here we only briefly review the fundamental aspects of BOP necessary for understanding of the model. In transition metals the *d*-states have energies comparable to the valence *s*-states. Owing to greater angular momentum, *d*-electrons do not extend so far from the nucleus. As a result, the wave functions of the *d*-states are quite localized in comparison with the *s*-states, and the behaviour of the *d*-electrons is intermediate between that of the valence and core electrons. Since the *d*-orbitals are constrained, they overlap only slightly with orbitals on neighbouring atoms and it is therefore natural to describe them within the tight binding theory rather than the free electron model.

In practice, the densities of states of the transition metals display a structure that is characteristic of a given crystal lattice. This structure is mainly determined by the interference of the *d*-orbitals and their mutual orientations given by the particular atomic arrangement. For *d*-orbitals, it is possible to form three types of bonds for which the angular momentum about the bond axis is preserved. In accordance with the molecular orbital theory these bonds are called *sigma, pi,* and *delta*, and their strengths are characterized by three bond integrals, labelled as *ddσ, ddπ* and *ddδ*. Among them the σ bond is strongest, since the lobes of the *d*-orbitals point towards each other and overlap most, followed by the π bond where the overlap is still significant. The δ bond is the weakest of the three, because the lobes of the *d*-orbitals are parallel to each other with only minimal overlapping. The ratios of the corresponding bond integrals computed within the canonical band theory [146] are as follows:

$$dd\sigma : dd\pi : dd\delta = (-6) : (+4) : (-1) \tag{2-1}$$

Based on the first-principles calculations it was shown that the *d*-bond integrals scale roughly as the inverse fifth power of the bond length [99]. The angular variations of these integrals as one atom is rotated around the other is of fundamental importance in understanding the angular dependence of bonding as discussed in [99]. The actual variations for the *d–d* interactions can then be specified in terms of directional cosines as the bond axis varies [147]. The detailed description of the *d*-bond integrals in the bond-order potential formalism can be found in the literature (e.g., [98, 148-150]). Within BOP, the binding energy of iron can be written as:

$$E_{binding} = E_{bond} + E_{rep} + E_{mag} \tag{2-2}$$

where E_{bond} is the attractive bond energy, E_{rep} is a repulsive term representing electrostatic and overlap repulsions and E_{mag} is the magnetic contribution obtained according to the Stoner model of itinerant magnetism [151-153].

The most important quantities determining E_{bond} in Eq. 2-2 are the two-center bond integrals $dd\sigma$, $dd\pi$, and $dd\delta$. Their distance dependence is represented by a continuous analytical function $\beta_\tau(r_{ij})$ that takes the generalized Goodwin-Skinner-Pettifor (GSP) form [154]:

$$\beta_\tau(r_{ij}) = \beta_\tau(r_0)(\frac{r_{ij}}{r_0})^{n_a} \exp\{n_b[(\frac{r_0}{r_c})^{n_c} - (\frac{r_{ij}}{r_c})^{n_c}]\} \qquad (2\text{-}3)$$

where r_{ij} is the distance between atoms i and j, r_0 the equilibrium separation of first nearest neighbors, and n_a, n_b, n_c and r_c are parameters determined directly from first-principles calculations [155]. The angular dependence of the intersite Hamiltonian matrix elements takes the usual Slater-Koster form [147].

Table 2-1. Fundamental properties of the ground-state ferromagnetic bcc iron used for fitting of BOP: lattice parameter a_0 [Å], cohesive energy per atom [eV], and elastic moduli [GPa].

a_0	E_{coh}	C_{11}	C_{12}	C_{44}
2.85	4.40	243.3	145.0	119.0

The magnetic contribution E_{mag} in Eq. 2-2 is crucial for correct description of magnetic iron phases. The Stoner model [151-153] introduces magnetism by including the presence of local exchange fields within the band energy. A comparison of densities of states (DOS) shows a good agreement between k-space TB and BOP for bcc and fcc phases of iron [100]. The accurate evaluation of the local DOS is necessary for correct determination of the magnetic energy that governs the relative stability of different magnetic bulk phases. The physically based description of bonding and magnetism is also crucial for the behaviour of lattice defects such

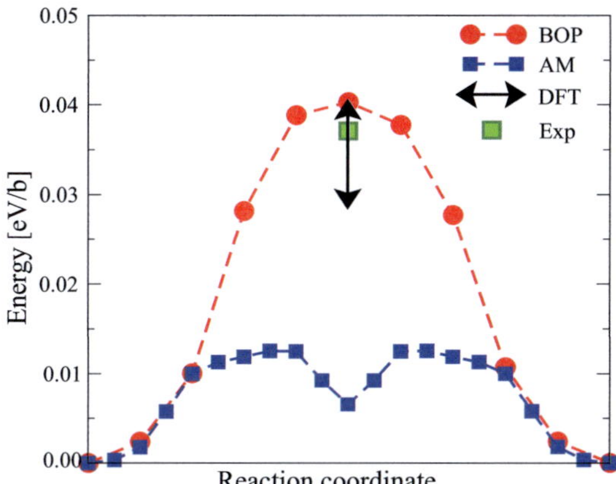

Figure 2-1. Peierls barriers for the straight $a_0/2\langle111\rangle$ screw dislocation moving between two neighbouring sites calculated using BOP, FS potential (AM) and DFT together with an experimental estimation (Exp) [100].

as dislocations that induce changes in bond lengths, bond angles and local magnetic moments [100]. The simple pair repulsive term in Eq. (2-2) is fitted to reproduce the fundamental properties of the equilibrium bcc ferromagnetic Fe (Table 2-1). The magnetic BOP for Fe has been shown [100] to reproduce correctly properties of ideal bulk phases as well as various crystal defects including dislocations.

In Fig. 2-1 it presents the comparison of the Peierls barriers for the $a_0/2\langle111\rangle$ screw dislocation moving between two neighbouring equilibrium lattice positions calculated using BOP, FS potential of Ackland and Mendelev (AM) [90, 156], and DFT together with an experimental estimation [100]. The AM potential is the only empirical potential that yields the non-degenerate core structure for iron. However, as shown in Fig 2-1, it predicts a double-hump shape of the energy barrier with a meta-stable dislocation configuration in the middle, which is unphysical. In contrast to the AM potential, the BOP energy barrier contains a single maximum located in the half-way position that agrees both qualitatively and quantitatively with the DFT calculations [157] and the experimental estimation [158]. Although the minimum energy path obtained from the NEB calculation can only be used for an estimation of the Peierls stress, it correctly reflects the lattice resistance to dislocation motion. As shown in Fig. 2-1, the agreement of the energy barriers between BOP and DFT/experimental results demonstrates its accuracy and reliability in predicting dislocation behaviour.

2.2   Nudged elastic band method

As mentioned in Chapter 1 the activation energy of the kink-pair formation can be determined by the integration of Eq. 1-9 in terms of the reaction coordinate along which the transformation of the dislocation core takes place. The nudged elastic band method is extensively used in our calculations to determine such transformations and in the following the method will be briefly summarized.

The phase transition is normally defined as a geometric and topological transformation process of a system from one phase to another, each of which has a unique and homogeneous physical property. The most important step involved in studying the phase transition is the knowledge of the activation energy barrier and the rate constant. In 1931, Erying and Polanyi proposed the transition state theory (TST) in terms of the activation energy and rate constants for characterizing reactions [159, 160]. In order to simulate a reaction or transition, a potential energy surface (PES) that characterizes the process is first generated. Then, a minimum energy path (MEP) is computed which represents the transition pathway in the reaction coordinate space. Finally, the activation energy and the rate constant that define the speed of the process can be calculated using TST.

A major challenge in searching MEP is the generation of the potential energy surface accurately. Reference [161] provides a detailed review of available methods to generate the PES characterizing information regarding the interatomic and intermolecular interactions that characterize the reaction. The MEP can be interpreted as the steepest descent path on the PES connecting the reactant and the product [162]. An important property of the MEP is that the direction of the gradient of the potential energy at any point on the MEP is the tangent direction along the MEP at that point. At the same time, for any degree of freedom perpendicular to the MEP at that point, the gradient of the potential energy is zero [162, 163]. Mathematically speaking, on the potential energy surface, the transition state is the first-order saddle point located between the local minima, i.e. the reactant and product along the MEP. Once the MEP is generated, the saddle point can be extrapolated. Then, using the transition state theory, one can estimate the activation energy and the transition rate constant. Various numerical methods to search transition paths and saddle points have been developed in the recent years (see [164-166] for review). Among them, the Nudged Elastic Band (NEB) method [130] and its improvements [128, 129, 167, 168] have become widely used due to their relative simplicity and easy implementation.

Here we briefly introduce this technique. The NEB method requires that the initial and final states are known. A number of intermediate states, usually between four and thirty, are iteratively adjusted and finally converge to the MEP keeping the initial and final state fixed. In general, the transition path is described by a set of $P+1$ images in configuration space with reactive coordinates:

$$\mathbf{R} = [\mathbf{R}_0, \mathbf{R}_1, \mathbf{R}_2, \cdots \mathbf{R}_P] \qquad (2\text{-}4)$$

Images are connected by an imaginary elastic band. The target MEP is a group of images where the total forces acting on them reach equilibrium, i.e., for any degree of freedom perpendicular to the MEP the energy is stationary.

The force acting on each image is a combination of the perpendicular component of the true force from the potential energy and the parallel component of the spring force projected along the unit tangent vector to the path. The force acting on image i is given by:

$$\mathbf{F}_i = -\nabla V(\mathbf{R}_i)\big|_{\perp} + \mathbf{F}_i^{s}\big|_{\parallel} \qquad (2\text{-}5)$$

The perpendicular component of the true force is written as:

$$-\nabla V(\mathbf{R}_i)\big|_{\perp} = \nabla V(\mathbf{R}_i) - \nabla V(\mathbf{R}_i) \cdot \hat{\boldsymbol{\tau}}_i \qquad (2\text{-}6)$$

The parallel components of spring force can be expressed as:

$$\mathbf{F}_i^{s}\big|_{\parallel} = k(|\mathbf{R}_{i+1} - \mathbf{R}_i| - |\mathbf{R}_i - \mathbf{R}_{i-1}|) \cdot \hat{\boldsymbol{\tau}}_i \qquad (2\text{-}7)$$

where V is the potential energy of the system and k is the spring constant. The tangent vector $\boldsymbol{\tau}_i$ is determined by the coordination of the neighboring images $\mathbf{R}_{i-1}$, $\mathbf{R}_i$ and $\mathbf{R}_{i+1}$.

To reduce the kinks in the MEP, only the adjacent image with higher energy is used in computing the tangent, unless i is at a maximum or a minimum. The tangent vector is calculated as following:

$$\boldsymbol{\tau}_i = \begin{cases} \boldsymbol{\tau}_i^+ \to V_{i+1} > V_i > V_{i-1} \\ \boldsymbol{\tau}_i^- \to V_{i+1} < V_i < V_{i-1} \end{cases} \qquad (2\text{-}8)$$

in which

$$\begin{cases} \boldsymbol{\tau}_i^+ = \mathbf{R}_{i+1} - \mathbf{R}_i \\ \boldsymbol{\tau}_i^- = \mathbf{R}_i - \mathbf{R}_{i-1} \end{cases} \qquad (2\text{-}9)$$

If the image i is at a maximum or a minimum the tangent vector is calculated based on a weighted average from the energy differences as following:

$$\boldsymbol{\tau}_i = \begin{cases} \boldsymbol{\tau}_i^+ \Delta V_i^{max} + \boldsymbol{\tau}_i^- \Delta V_i^{min} \to V_{i+1} > V_{i-1} \\ \boldsymbol{\tau}_i^+ \Delta V_i^{min} + \boldsymbol{\tau}_i^- \Delta V_i^{max} \to V_{i+1} < V_{i-1} \end{cases} \qquad (2\text{-}10)$$

where

$$\begin{cases} \Delta V_i^{max} = \max(|V_{i+1} - V_i|, |V_{i-1} - V_i|) \\ \Delta V_i^{min} = \min(|V_{i+1} - V_i|, |V_{i-1} - V_i|) \end{cases} \qquad (2\text{-}11)$$

With the determined forces both parallel and perpendicular to the tangent, the elastic band can be relaxed using any optimization algorithm. At each iteration, the forces acting on all images are minimized at the same time. As a result, the whole elastic band iteratively converges to the MEP.

2.3   Simulation geometry

The atomic simulation block used in our calculations is depicted schematically in Fig. 2-2 and its main characteristics are given below. The rectangular block is periodic along the z-direction, which coincides with the direction of the [111] dislocation line. The periodic length of the block in the z direction equals to the Burgers vector $b = \sqrt{3}/2a_0$ ($a_0 = 2.85\,\text{Å}$, the

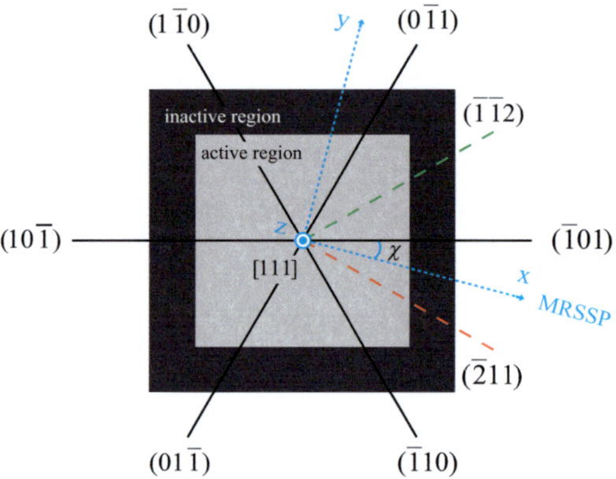

Figure 2-2. Simulation block used in the atomistic calculations. The $a_0/2<111>$ screw dislocation is introduced in the center according to the anisotropic elastic displacement. The inactive region (black) extends effectively to infinity. The orientation of the MRSSP is defined by the angle χ.

lattice parameter of Fe). Therefore, the dislocation in our simulations is always straight and infinite without any kinks or jogs. The y-axis is perpendicular to the $(\bar{1}01)$ plane, and the x-axis is perpendicular to the $(\bar{1}2\bar{1})$ plane. In the x and y directions perpendicular to the dislocation line rigid boundary conditions are used. In this setup, the atoms in the outmost 'inactive' region are kept fixed so that the dislocation is effectively placed in an infinite crystal environment. The dimensions of the block in the x and y directions are about 20×20 lattice parameters, which is large enough for a single screw dislocation. The $a_0/2<111>$ screw dislocation is introduced in the center of the perfect lattice by displacing the atoms according to the anisotropic elastic displacement field in an infinite medium [52]. In order to obtain realistic core structures, the atomic positions in the active region of the block are fully relaxed while those in the inactive region are fixed to maintain the infinite elasticity field of the screw dislocation. The relaxation is considered complete when the forces on all atoms fall below 0.001 eV/Å. In all our calculations, an efficient relaxation algorithm, the fast inertial relaxation engine (FIRE) [169], was implemented.

3 Results

3.1 Atomistic study of the *1/2*<111> screw dislocation

As mentioned in Chapter 1, the most prominent features of deformation behavior of bcc metals and alloys are the breakdown of the Schmid law [42-47], and the rapid increase of the yield and flow stresses with decreasing temperature and increasing strain rate. These macroscopic mechanical properties are consequences of intrinsic properties of the $a_0/2$<111> screw dislocations at the atomic scale [52-56].

Fig. 3-1 shows the relaxed core structure of the $a_0/2$<111> screw dislocation in iron computed using the magnetic BOP. The atomic arrangements are shown using the differential displacement map [81, 84, 170] in the planes perpendicular to the dislocation line. The lengths of the arrows connecting atoms correspond to the relative displacements of two neighboring atoms parallel to the Burgers vector. Each of the three longest arrows in the center of the figure corresponds to a relative displacement equal to $1/3b$, defining a circuit that gives a complete Burgers vector of $1/2[111]$ of the dislocation. The same net product can be also obtained for any other circuit going around the center of the screw dislocation. The

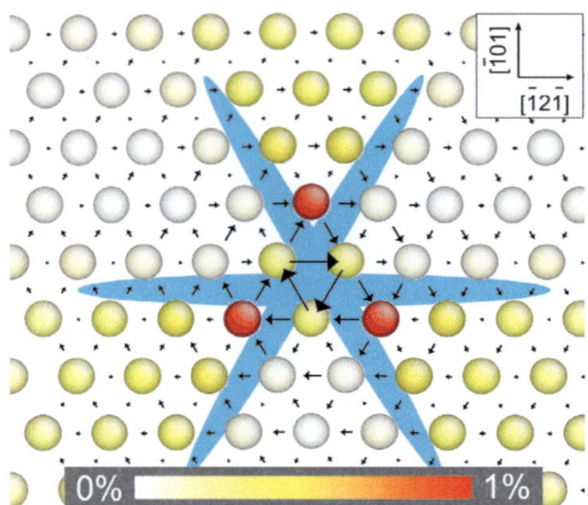

Figure 3-1. Core structure of the $a_0/2[111]$ screw dislocation. The arrows are the displacement of two neighboring atoms in the [111] direction parallel to the Burgers vector relative to their separation in the perfect lattice. The blue shading is used to highlight the form of the core spreading. The coloring of the atoms shows the relative decrease of atomic magnetic moments from their bulk value.

core structure is non-degenerate, extended symmetrically on three {110} planes, virtually identical to that found in DFT calculations [93-97].

In addition to the core structure, the magnetic BOP can also reveal the changes of the local magnetic moments of Fe atoms. For the $a_0/2\langle111\rangle$ screw dislocation these changes are relatively small, since the main structural changes occur in bond angles rather than in bond lengths. The screw dislocation therefore possesses the least distorted core structure relative to the perfect lattice among all dislocations. In contrast to the screw dislocation, much larger distortions and much larger changes (decreases) of the magnetic moments were found at the cores of $a_0/2\langle111\rangle$ edge and [100] dislocations [100].

In this section the mechanical response of the 1/2[111] screw dislocation to a series of external loadings will be examined. The purpose is to identify the stress components that affect the motion of individual screw dislocation and subsequently to quantify their effects on the magnitude of the Peierls stress. To study these phenomena at the atomic level, the following loadings were applied to the simulation block containing a 1/2[111] screw dislocation in its center:

a) a set of pure shear stresses parallel to the slip direction in different maximum resolved shear stress planes, to verify the twinning-antitwinning effect;

b) a set of uniaxial loadings in tension and compression with MRSSPs corresponding to those in **a)** to examine the tension-compression asymmetry;

c) a set of combined stress tensors with stress components both parallel and perpendicular to the slip direction to quantitatively determine the effects of the stress components on the motion of the dislocation.

In all calculations, the applied stresses were superimposed on the simulation block by displacing atoms in both the active and inactive regions according to the corresponding strain field determined using anisotropic elasticity theory [52]. The atoms in the active region were then fully relaxed while those in the inactive region were kept fixed to maintain the applied stress. The applied stress was increased gradually and full relaxation was carried out after every stress increment until the dislocation started to move.

3.1.1 Loading by pure shear stress parallel to the slip direction

The loadings with pure shear stress, which cannot be easily applied in experiments, are the simplest and most direct measurements of the Peierls stress that reveal whether the material behaves according to the Schmid law. The loading geometries are illustrated in Fig. 2-2. The orientation of the MRSSP is defined by the angle χ that it makes with the $(\bar{1}01)$ plane. The pure shear stress parallel to the slip direction was applied in different MRSSPs using the following stress tensor:

$$\Sigma_\sigma = \begin{bmatrix} 0 & 0 & 0 \\ 0 & 0 & \sigma \\ 0 & \sigma & 0 \end{bmatrix} \tag{3-1}$$

Σ_σ is defined in the right-handed coordinate system with the y-axis normal to the MRSSP, the z-axis parallel to the [111] direction and the x-axis lying in the MRSSP. The shear stress σ was built up incrementally in steps of $0.0005C_{44}$, where C_{44} is the elastic modulus. When the resolved shear stress in the MRSSP reached the critical value (CRSS), the dislocation started to glide. This "visual" determination of CRSS corresponds to a discontinuity on the energy-stress plot, so that an accurate value CRSS can be determined unequivocally from the simulations. In order to investigate the twinning-antitwinning asymmetry, a set of MRSSPs are chosen as illustrated in Fig. 3-2. Owing to the lattice symmetry, it is only necessary to consider $-30° \leq \chi \leq 30°$.

The resulting dependence of the CRSS on the orientation of the MRSSP is plotted in Fig. 3-3. For all loadings with pure shear stress parallel to the

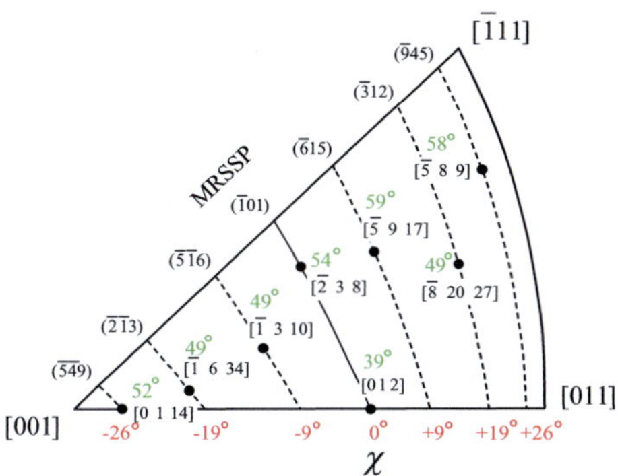

Figure 3-2. Standard stereographic triangle for which the ($\bar{1}$01) plane is the most highly stressed {110} plane in the [111] zone. Uniaxial loading orientations are in square brackets and their corresponding MRSSPs are in parenthesis. The angle χ is colored by red along the [001]-[011] side and λ by green.

slip direction, the observed glide plane is the ($\bar{1}$01) plane. The Schmid law is plotted in the figure as dashed line. If the Schmid law applies, the projection of the CRSS onto the ($\bar{1}$01) plane should be the same for any orientation of the MRSSP and the CRSS should be proportional to $\cos^{-1}\chi$ as plotted in Fig. 3-3. Not surprisingly, the calculated CRSS-χ dependence for iron evidently deviates from the Schmid law. The orientations with positive and negative χ are not equivalent, and the CRSS is higher for positive than for negative χ values. This is the well-known twinning-antitwinning asymmetry observed in bcc metals both in experiments (e.g. [44, 54]) and in other atomistic calculations (e.g. [101]). In our calculations with positive σ, the pure shear stresses with $0° \leq \chi \leq 30°$ bounded by ($\bar{2}$11) and ($\bar{1}$01) planes are in the antitwinning sense while those with $-30° \leq \chi \leq 0°$ bounded by ($\bar{1}\bar{1}$2) and ($\bar{1}$01) are in the twinning sense.

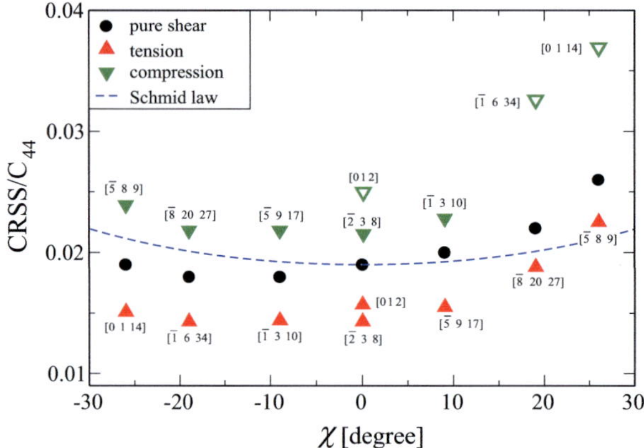

Figure 3-3. Dependence of the CRSS on the orientation of the MRSSP, χ, for loadings by pure shear stress parallel to the slip direction in the MRSSP (circles), in tension (up-triangles) and compression (down-triangles).

3.1.2   Loading in tension and compression

Experimentally, iron shows not only the twinning-antitwinning asymmetry but also the tension-compression asymmetry, which is again the consequence of the non-planar core structure of the $a_0/2{<}111{>}$ screw dislocation. In simulations, the tension and compression loadings are important tests, since they can disclose whether or not the shear stress parallel to the slip direction is the only stress component that affects the dislocation motion. Such simulations can be also directly compared to experimental results as most experiments are performed under uniaxial loadings.

The loading orientations for tension investigated in our study are shown in Fig. 3-2 as solid circles. The MRSSP of each of these uniaxial loadings corresponds to the MRSSP which has been used in the previous section for the pure shear loadings with only shear stress parallel to the slip direction. One should note that with the same loading direction the signs of the shear stress components parallel to the slip direction are reversed for tension and compression. It means that when comparing the compression results to those of tension, the sign of χ should also be reversed. The reason is the twinning-antitwinning asymmetry: the CRSS depends on the sense of the shearing so that changing the sign of the shear stress components parallel to the slip direction is equivalent to changing the sign of χ while keeping the shear stress fixed.

The critical value of the uniaxial loadings projected on the MRSSP parallel to the [111] direction can be directly compared to the CRSS determined with pure shear stress parallel to the slip direction in the last section. The purpose is to verify whether the CRSSs are the same for the pure shear loadings and the uniaxial loadings with the same χ. If they are the same - no other stress component affects the CRSS; if they are not the same, other components of the stress tensor influencing the Peierls stress can be qualitatively determined. Such analysis was first performed by Ito and Vitek [89] for molybdenum and tantalum using the FS potential, and more recently by Gröger et al. [101] for tungsten and molybdenum using BOP. Both studies confirmed that the shear stress perpendicular to the slip direction indeed strongly affected the glide of the screw dislocation in bcc metals.

Our results for Fe, displayed in Fig. 3-3, show that for a given χ the CRSS for compression is always considerably higher than that for tension and the CRSS for the pure shear loading lies in between. This confirms that it's the same for iron, the glide of the screw dislocation is significantly affected by other stress components, which make the screw dislocation

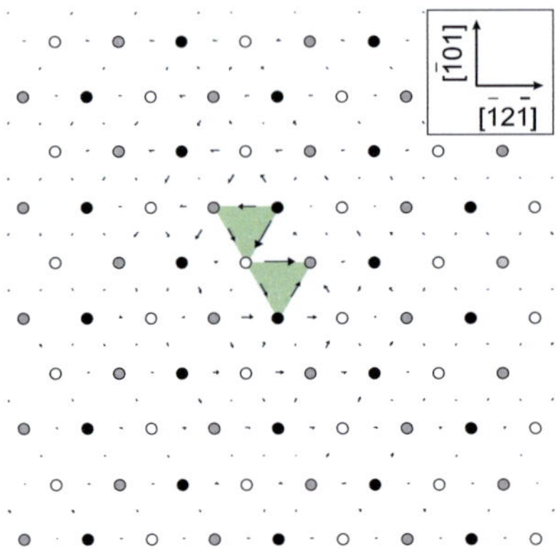

Figure 3-4. Differential displacement plot for the slip core structure, induced by high non-Schmid stress components in compression. The shading is used to highlight the distribution of the Burgers vector.

either easier to glide (in tension) or harder (in compression). For all loadings in tension the screw dislocation glided on the $(\bar{1}01)$ plane, which is also the glide plane for loadings with pure shear stress parallel to the slip direction. For most loadings in compression (the solid down-triangles in Fig. 3-3), the glide plane changes from $(10\bar{1})$ to $(1\bar{1}0)$, although the Schmid factor is two times higher on the former slip system. This indicates that the shear stress perpendicular to the slip direction strongly affects the dislocation core and makes its glide easier on the inclined {110} planes of the [111] zone.

The empty triangles in Fig. 3-3 present special cases. With compression in [012] direction, the first jump of the screw dislocation was not a single elementary step on a {110} plane but a multi-step motion, seemingly on the $(2\bar{1}\bar{1})$ plane. A careful analysis of the atomic structures however showed that such glide is a combination of alternating jumps on two

{110} glide planes. A calculation with smaller increment of the loading step confirmed that two glide planes, $(10\bar{1})$ and $(1\bar{1}0)$, were activated one after the other, so that the elementary motion is always of {110} type and no direct jump of the screw dislocation along the $(2\bar{1}\bar{1})$ plane occurs.

For loadings in compression with $\chi = +19°$ or $+26°$ the dislocation core structure markedly changes at the applied stress of about $0.05C_{44}$. The large non-Schmid stresses cause the core to extend significantly on the inclined $(1\bar{1}0)$ plane (Fig. 3-4). The CRSS to move this extended core is so high that other slip systems will be activated before such stress level is reached (see later).

The CRSS for uniaxial loadings with various directions in Fig. 3-3 confirmed the experimentally observed tension-compression asymmetry [44]. Most importantly, these simulations show that the shear stress components other than those parallel to the slip direction significantly affect the glide of the screw dislocation. In the following, this aspect of dislocation behavior will be investigated in detail, using loadings with both shear stress parallel and perpendicular to the slip direction.

3.1.3   Loading by shear stress perpendicular to the slip direction combined with shear stress parallel to the slip direction

Previous atomistic studies on bcc metals [53, 89, 101, 170] have shown that the influences on the CRSS by the shear stress perpendicular to the slip direction are related to changes of the core structure induced by these non-Schmid stresses. For example, the change of the dislocation core in bcc molybdenum under shear stress perpendicular to the slip direction with $\chi = 0$ was examined in [101]. The authors showed, that the core either extends or constricts on the glide planes in the [111] zone, depending on the magnitude and direction of the non-Schmid stresses. As a consequence, the motion of the dislocation was either promoted or suppressed, e.g., the CRSS will increase if the core is constricted on the glide plane and, vice versa, it will decease when the core extends along the glide plane. The same explanation applies also for iron, and it will be discussed in detail in Chapter 4.

In order to investigate the dependence of the CRSS on the magnitude of the shear stress, τ, perpendicular to the slip direction, a set of simulations of the $a_0/2[111]$ screw dislocation, subjected to simultaneous loading by various combination of shear stresses both parallel and perpendicular to the slip direction, were carried out. The specific stress tensor applied is:

$$\Sigma_{\sigma,\tau} = \begin{bmatrix} -\tau & 0 & 0 \\ 0 & \tau & \sigma \\ 0 & \sigma & 0 \end{bmatrix} \tag{3-2}$$

This stress tensor is similar to that in Eq. 3-1, except of the presence of the additional diagonal stress components $\pm\tau$, which represent the shear stresses perpendicular to the [111] direction in the coordinate system rotated by $-45°$ around the z-axis.

By reducing the stress tensor, e.g. of uniaxial loadings, into the format of Eq. 3-2 with only shear stresses parallel and perpendicular to the slip direction, it can be verified that only the shear stresses parallel and perpendicular to the slip direction determine the motion of the dislocation and, all other non-zero stress components, e.g. hydrostatic stress [171], have no effect on the dislocation motion. If this is indeed so, the results should be identical to (or at least very similar to) the CRSS-τ relationship obtained for the uniaxial loadings in tension and compression described in the previous section.

The application of the combined stress tensor in Eq. 3-2 was again done in several steps. First, for a given χ the shear stress τ perpendicular to the slip direction was superimposed on the block according to the elasticity theory. Then the shear stress σ was built up incrementally in steps of $0.0005C_{44}$, until the resolved shear stress in the MRSSP reached the CRSS and consequently the dislocation started to move. By repeating this process with different values of τ, the dependencies of the CRSS on the shear stress perpendicular to the slip direction are obtained for χ values corresponding to the orientations investigated in the uniaxial loadings.

Fig. 3-5 shows the dependences of the CRSS on τ for five MRSSP orientations, which are the $(\overline{2}\,\overline{1}3)$ plane with $\chi=-19°$, the $(\overline{5}\,\overline{1}6)$ plane with $\chi=-9°$, the $(\overline{1}01)$ plane with $\chi=0°$, the $(\overline{6}15)$ plane with $\chi=+9°$, and the $(\overline{3}12)$ plane with $\chi=+19°$. For all of these orientations, the CRSS is lower for positive τ than that of $\tau=0$, and in this region the dislocation always glides on the $(\overline{1}01)$ plane. In contrast, negative τ makes the glide on the $(\overline{1}01)$ plane more difficult. In the region of negative τ, the CRSS increases with decreasing τ until the glide plane changes from the $(\overline{1}01)$ plane to the $(0\,\overline{1}1)$ plane. The transitions of the glide plane are indicated by the arrows in the bottom of Fig. 3-5. When $\tau \approx -0.01C_{44}$, two slip systems, $(\overline{1}01)[111]$ and $(0\,\overline{1}1)[111]$, become activated for orientations corresponding to $\chi=0°$, $-9°$, and $-19°$. For these loadings, the motion of the screw dislocation is therefore expected to be composed of alternating elementary jumps on neighboring {110} planes, leading to macroscopic glide along the $(\bar{1}\,\bar{1}2)$ plane.

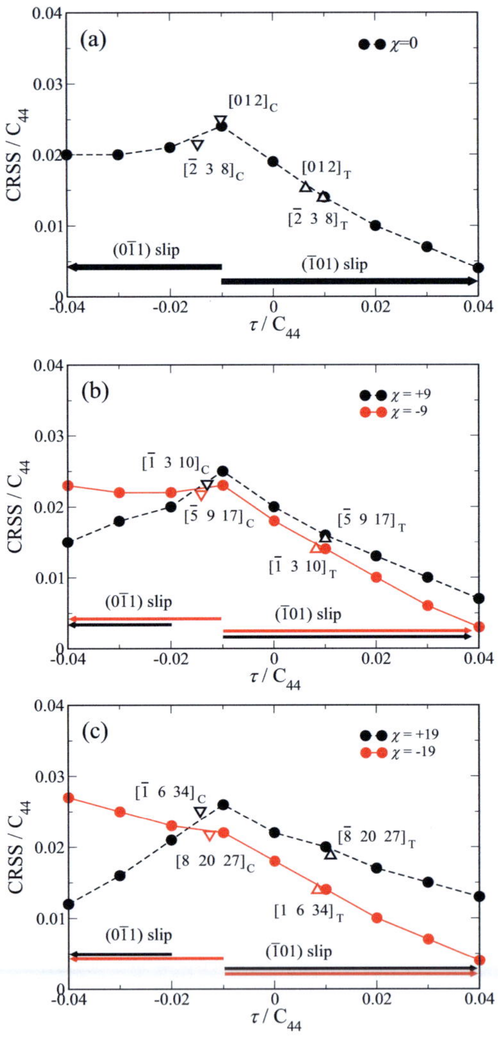

Figure 3-5. Dependence of the CRSS on the shear stress perpendicular to the slip direction, τ, for different MRSSPs with (a) $\chi=0°$, (b) $\chi=\pm9°$, and (c) $\chi=\pm19°$. Triangles correspond to the uniaxial loadings.

Fig. 3-5 also contains the CRSS-τ results, labeled as triangles, from the corresponding uniaxial loading simulations described in Section 3.1.2 (cf. Fig. 3-3). The shear stress τ perpendicular to the slip direction and its corresponding CRSS were extracted from the general stress tensor of uniaxial loadings. It should be noted that only the deviatoric part of the stress tensor should be taken into account when evaluating the CRSS and τ, since the hydrostatic stresses do not affect the dislocation motion [171]. Then the results from uniaxial loadings can be directly compared with the CRSS-τ dependence obtained using the combined stress tensor in Eq. 3-2 with shear stresses both parallel and perpendicular to the slip direction. It can be seen from Fig. 3-5 that there is a good agreement between them. This agreement confirms that for any loading only two shear stress components, those parallel and perpendicular to the slip direction, determine the motion of the $a_0/2<111>$ screw dislocation in Fe.

As a brief conclusion for the static atomistic studies, our simulations show that the CRSS of the $a_0/2[111]$ screw dislocation depends only on two factors: the orientation of the slip system given by χ and the non-Schmid shear stresses given by τ. The physical origins of these effects will be discussed in detail in Chapter 4.1.

3.2   Yield criterion for single crystal

Despite the fact that the onset of the plastic deformation is determined by properties of single dislocations, engineering calculations are based on continuum yield criterions that represent the microscopic behavior by few fundamental parameters. The early framework of the continuum description for single crystal plasticity was developed by Hill [103] and Rice [104]. These theories are commonly based on the Schmid law for close-packed fcc and hcp metals. However, it was shown by atomistic simulations in the last section that for iron the non-Schmid stress, i.e. the stress components perpendicular to the slip direction which do not drive the dislocation glide in the slip plane, also affect the CRSS. This indicates the common continuum model assuming Schmid-type plastic behaviour is not well suited for bcc iron. Thus, in the current section an appropriate yield criterion will be formulated that accommodates to the effects of both the shear stresses parallel and perpendicular to the slip direction. This is done following the works of Qin and Bassani [107, 108] and Gröger [109]. As the next step, the constructed yield criterion will be employed to determine the yield surface and compare the results to those obtained with Schmid law.

3.2.1   24 slip systems in bcc metals

Owing to the lattice symmetry the atomistic results obtained in the last section are also applicable for the other two glide planes, i.e. the $(01\bar{1})$ and $(1\bar{1}0)$ planes, in the [111] zone if one rotates the coordinate system and the loading around the [111] direction (z-axis) by $\pm 2\pi/3$. Furthermore, in any bcc crystals there are four equivalent $\{111\}$ directions and in each of them three independent $\{110\}$ glide planes exist [52, 54]. In addition, the positive and negative slip directions need to be distinguished due to the twinning-antitwinning asymmetry [44, 53]. A convenient way to capture this effect is to change the sign of χ while keeping the sense of the shear fixed. This increases the number of {111} directions from four to eight. Thus, there are in total 24 {110}<111> reference systems, in which only loadings with positive shear stress parallel to the slip direction need to be considered. The complete list of these slip systems was given in [109] and is presented it in Table 3-1 for later use. Note that the systems 13 to 24 are conjugate to the systems 1 to 12. A pair of systems α and $\alpha + 12$ have identical glide plane but opposite slip direction.

Table 3-1. The 24 slip systems in bcc crystals.

α	$(\mathbf{n}^\alpha)[\mathbf{m}^\alpha]$	$[\mathbf{n}_1^\alpha]$	α	$(\mathbf{n}^\alpha)[\mathbf{m}^\alpha]$	$[\mathbf{n}_1^\alpha]$
1	$(01\bar{1})[111]$	$[\bar{1}10]$	13	$(01\bar{1})[\bar{1}\bar{1}\bar{1}]$	$[10\bar{1}]$
2	$(\bar{1}01)[111]$	$[0\bar{1}1]$	14	$(\bar{1}01)[\bar{1}\bar{1}\bar{1}]$	$[\bar{1}10]$
3	$(1\bar{1}0)[111]$	$[10\bar{1}]$	15	$(1\bar{1}0)[\bar{1}\bar{1}\bar{1}]$	$[0\bar{1}1]$
4	$(\bar{1}0\bar{1})[\bar{1}11]$	$[\bar{1}\bar{1}0]$	16	$(\bar{1}0\bar{1})[1\bar{1}\bar{1}]$	$[01\bar{1}]$
5	$(0\bar{1}1)[\bar{1}11]$	$[101]$	17	$(0\bar{1}1)[1\bar{1}\bar{1}]$	$[\bar{1}\bar{1}0]$
6	$(110)[\bar{1}11]$	$[01\bar{1}]$	18	$(110)[1\bar{1}\bar{1}]$	$[101]$
7	$(0\bar{1}\bar{1})[\bar{1}\bar{1}1]$	$[1\bar{1}0]$	19	$(0\bar{1}\bar{1})[11\bar{1}]$	$[\bar{1}0\bar{1}]$
8	$(101)[\bar{1}\bar{1}1]$	$[011]$	20	$(101)[11\bar{1}]$	$[1\bar{1}0]$
9	$(\bar{1}10)[\bar{1}\bar{1}1]$	$[\bar{1}0\bar{1}]$	21	$(\bar{1}10)[11\bar{1}]$	$[011]$
10	$(10\bar{1})[1\bar{1}1]$	$[110]$	22	$(10\bar{1})[\bar{1}1\bar{1}]$	$[0\bar{1}\bar{1}]$
11	$(011)[1\bar{1}1]$	$[\bar{1}01]$	23	$(011)[\bar{1}1\bar{1}]$	$[110]$
12	$(\bar{1}\bar{1}0)[1\bar{1}1]$	$[0\bar{1}\bar{1}]$	24	$(\bar{1}\bar{1}0)[\bar{1}1\bar{1}]$	$[\bar{1}01]$

The derivation of the yield stress for iron single crystal is performed in the following way: first a particular {110} reference plane is defined in the zone of the slip direction from which the angle of the MRSSP, χ, and the angle of the slip plane, ψ, can be measured. Each of the 24 reference systems defined by a reference plane and a slip direction can be determined in the same way. If one neglects the interactions between dislocations, the motion of each individual dislocation is governed by the same $CRSS - \chi$ and $CRSS - \tau$ dependencies obtained for the isolated $a_0/2[111]$ dislocation from the atomistic simulations. To apply these dependencies to any reference system, the angle χ_α of the MRSSP in the zone of the corresponding <111> slip direction is required to lie within the ±30° angular region measured from the respective {110} reference plane. Additionally, it is required that the shear stress parallel to the slip direction resolved in each of these MRSSP, σ_α, is positive. Consequently, there are only 4 out of the total 24 reference systems satisfying all requirements, which can be activated for slip by the applied stress. For the opposite sense of loading, the four reference systems are sheared in the opposite sense and thus the 4 slip systems that can be activated change to the conjugate ones (cf. Table. 3-1).

The atomistic study of the $a_0/2[111]$ screw dislocation in the last chapter indicated that the relationship between the CRSS and τ is unique for a given χ, independently of the loading history how the corresponding shear stresses σ and τ were attained. The dependences of CRSS on τ were achieved for a set of discrete χ values ($-26°$, $-19°$, $-9°$, $0°$, $+9°$, $+19°$ and $+26°$) in our studies (Fig. 3-5). For each of the four reference systems α, the shear stresses parallel and perpendicular to the slip direction, associated with a certain loading, can be determined as a stress pair, $(\sigma_\alpha, \tau_\alpha)$. Since all $a_0/2$<111> dislocations are equivalent, the shear stresses $(\sigma_\alpha, \tau_\alpha)$ of the four {110}<111> slip systems can now be directly compared with atomistic results obtained for the isolated $a_0/2[111]$ dislocation. One should note, in order to develop a complete description of the yielding of a single crystal, the simulations performed in Chapter 3.1.3 have to be repeated for all χ values between $-30°$ and $30°$. Then for any loading, the shear stress components parallel and perpendicular to the slip direction can be extracted into the corresponding MRSSP coordinate system and then used to determine the commencement of the dislocation motion in each of these 24 slip systems [109]. Here this process is demonstrated using a special loading for which our on-hand atomistic data can be used.

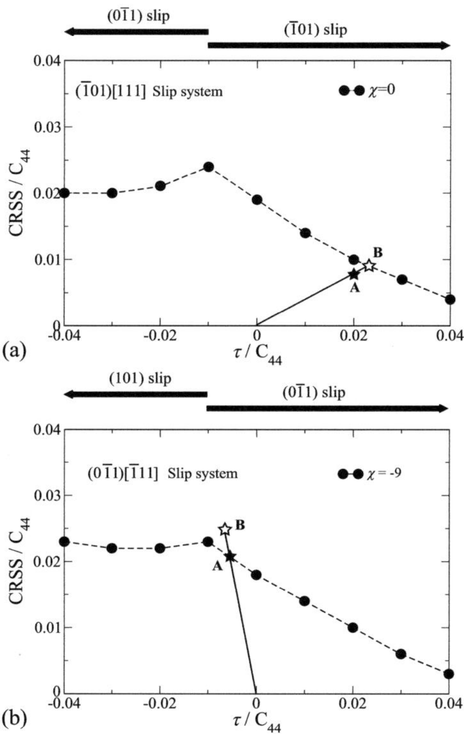

Figure 3-6. Evolution of the same loading in two different {110}<111> slip systems. Squares correspond to the atomistic data calculated for a single $a_0/2$[111] dislocation in Section 3.1.3. The points **A** and **B** in the two panels correspond to the yield point in the two slip systems.

Let us consider a reduced stress tensor in the format of Eq. 3-2, with both shear stresses parallel and perpendicular to the slip direction and $\chi = 0$, applied on the slip system $(\bar{1}01)[111]$ ($\alpha = 2$). The corresponding CRSS-τ dependence is shown in Fig. 3-5(a). Obviously, when τ is close to zero, $(\bar{1}01)[111]$ is the primary slip system. However, as the magnitude of τ increases the shear stresses, both parallel and perpendicular to the slip direction, evolve also in other slip systems. For a given loading, the loading path which defines a unique dependence of σ_α and τ_α, in each slip system is different, e.g., with unique $\eta_\alpha = \tau_\alpha / \sigma_\alpha$ (note η_α are usually not the same in different slip systems). For example in Fig. 3-6(a), the loading path with $\eta_2 = 2.26$ in the $(\bar{1}01)[111]$ slip system ($\alpha = 2$) is plotted as the straight line passing through the origin and extending towards the points representing the dependence of the CRSS on τ for $\chi = 0$. If only $a_0/2[111]$ dislocations were gliding, the $(\bar{1}01)[111]$ system would become operative at the point where this line intersects the CRSS-τ dependence, i.e. at the point marked as **B**. However, if dislocations with Burgers vector other than $[111]$ but, for example, $[\bar{1}11]$ are considered, while the loading in the $(\bar{1}01)[111]$ reference system proceeds along the path shown in Fig. 3-6(a), another reference system, $(0\bar{1}1)[\bar{1}11]$ ($\alpha = 5$), in which the orientation of the MRSSP corresponds to $\chi = -9°$, is subjected to the shear stress pairs, τ_5 and σ_5, that evolve along the loading path $\eta_5 = -0.19$. This path is shown as a straight line in Fig. 3-6(b) passing through the origin. The path was determined employing the procedure outlined previously using the reduced stress tensor translated into the corresponding MRSSP coordinate system in the format of Eq. 3-2. As already emphasized, the CRSS-τ dependence achieved in atomistic simulations is the same for every system

α. This means that the CRSS-τ dependence with $\chi = -9°$ for the 1/2[111] dislocation [see Fig. 3-5(b) in Chapter 3.1.3] can be directly applied to the 1/2[$\bar{1}$11] dislocation with the same MRSSP angle. The system $(0\bar{1}1)[\bar{1}11]$ becomes operative at the point where the loading path in Fig. 3-6(b) intersects the CRSS-τ dependence, i.e. at the point marked **A**. We see as the loading increasing from zero, it reaches first point **A**, which corresponds to the critical yield point in the slip system $\alpha = 5$. Consequently, the $(0\bar{1}1)[\bar{1}11]$ system will become operative before the $(\bar{1}01)[111]$ system, and it is thus the active slip system for the loading considered.

The above example applies only for the particularly designed loading, for which the previous atomistic data (Fig. 3-5) can be utilized. Nevertheless, it has to be mentioned that the analysis proves that the CRSS-τ dependences from the atomistic simulations can be used to estimate the yielding point for any type of loading. However, when considering an arbitrary loading, this procedure is extremely computational expensive since the CRSS-τ dependences for all χ values have to be established from the atomistic calculations. Instead, it is much more efficient and intelligent to formulate an analytical yield criterion that applies to all 24 {110}<111> slip systems and can reproduce with sufficient accuracy the achieved atomistic results.

The Schmid law is the simplest yield criterion, which is well established for plastic deformation of fcc and hcp metals. When only the shear stress component in the slip direction determines the yielding, the plastic flow is called associated and the criterion is virtually the Schmid law. However, if other stress components, which do not directly drive the dislocation glide in the slip plane, also affect the yielding and the plastic flow, the flow is called non-associated. To accommodate such non-Schmid behavior, Qin and Bassani [107, 108] proposed a generalized Schmid law in which the stress components other than the Schmid stress also enter the yield criterion. This generalization of the Schmid law showed its accuracy in predicting the tension-compression asymmetry and the orientation dependence of the CRSS observed in $L1_2$ intermetallic compounds. Following this work, a yield criterion employing two shear stresses parallel to the slip direction, resolved on two {110} planes in the same <111> zone, was used to reproduce the twinning-antitwinning asymmetry obtained from the atomistic studies using both central-force many-body potential and BOP for molybdenum [172, 173]. Based on these studies, Vitek [174] and Gröger [109] formulated a general form of yield criteria for the non-associated flow in bcc metals. In order to capture the dependences of the CRSS on both the loading orientation and on the stress components other than those parallel to the slip direction, the analytical yield criterion comprises two shear stresses parallel and two shear stresses perpendicular to the slip direction, both resolved in two different {110} planes of the [111] zone. This approach was illustrated in [109] and reproduced both the twinning-antitwinning and the tension-compression asymmetries observed experimentally and atomistically (for more details see [101, 109]).

In the following, the analytical yield criterion will be developed for the yielding and the plastic flow in iron which will reproduce closely the dependences of the CRSS on χ and τ obtained from the atomistic simulations for iron. This yield criterion can be employed to determine the yield surface projected onto the CRSS-τ plot and compared to the atomistic results to verify its accuracy. Furthermore, the results can be used to predict the active slip systems that operate for any loading conditions at 0K.

3.2.2　Construction of analytical yield criterion

From the atomistic results obtained in Chapter 3.1.3, one can see that for a given χ the CRSS depends to a good approximation linearly on τ. This indicates that a linear yield criterion can be formulated to reproduce the atomistic data. In 2008 Gröger et al. [109] formulated a general form of yield criterion for the non-associated flow in bcc metals. This general yield criterion is able to capture both the χ and τ dependence of the CRSS by including two shear stresses parallel and two shear stresses perpendicular to the slip direction, both resolved in two different $\{110\}$ planes of the $[111]$ zone. For the $a_0/2[111]$ screw dislocation, this general yield criterion is written as:

$$\sigma^{(\bar{1}01)} + a_1\sigma^{(0\bar{1}1)} + a_2\tau^{(\bar{1}01)} + a_3\tau^{(0\bar{1}1)} = \tau^*_{cr} \qquad (3\text{-}3)$$

where $\sigma^{\{110\}}$ and $\tau^{\{110\}}$ are the shear stresses parallel and perpendicular to the slip direction, resolved in the corresponding $\{110\}$ planes. One should note that the selection of the second $\{110\}$ plane, other than the $(\bar{1}01)$ primary glide plane, can be arbitrary. Different choices may lead to different fitting results, but the overall outcome of the criterion remains independent of this choice. The first term in Eq. 3-3 corresponds to the stress that drives the dislocation to move in the $(\bar{1}01)$ glide plane. By neglecting the remaining terms on the left, Eq. 3-3 simply reduces to the Schmid law:

$$\sigma^{(\bar{1}01)} = \tau^*_{cr} \qquad (3\text{-}4)$$

so that $\sigma^{(\bar{1}01)}$ is commonly called as the Schmid stress. In contrast, the other stresses $\sigma^{(0\bar{1}1)}$, $\tau^{(\bar{1}01)}$ and $\tau^{(0\bar{1}1)}$ in Eq. 3-3 affect the structure of the dislocation core but do not directly exert any driving force on the dislocation. These stresses are therefore referred as the non-Schmid stresses. The second term, $\sigma^{(0\bar{1}1)}$, is the shear stress parallel to the slip direction in the $(0\bar{1}1)$ plane and, together with $\sigma^{(\bar{1}01)}$, reproduces the effect of the twinning-antitwinning asymmetry on the CRSS. The yield criterion employing only the first two terms was employed earlier in [172, 173] and successfully reproduced the twinning-antitwinning asymmetry obtained from the results of the atomistic studies for molybdenum. The third and fourth terms are two shear stresses perpendicular to the slip direction in two $\{110\}$ planes.

The coefficients a_1, a_2, and a_3, as well as τ_{cr}^{*} in Eq. 3-3 are parameters which are determined by fitting the yield criterion to the dependences of CRSS on χ and τ obtained from atomistic calculations. Typically, a_1 and τ_{cr}^{*} are fitted first using the CRSS vs. χ dependence under loadings with pure shear stress parallel to the slip direction on different MRSSPs. In this case Eq. 3-3 reduces to:

$$\sigma^{(\bar{1}01)} + a_1\sigma^{(0\bar{1}1)} = \tau_{cr}^{*} \tag{3-5}$$

The two shear stresses in Eq. 3-5 can be written in terms of the CRSS and the MRSSP orientation χ:

$$\sigma^{(\bar{1}01)} = CRSS\cos\chi \tag{3-6}$$

and

$$\sigma^{(0\bar{1}1)} = CRSS\cos(\chi + \pi/3) \tag{3-7}$$

Then for a given χ the corresponding CRSS can be determined as:

$$CRSS(\chi) = \frac{\tau_{cr}^*}{\cos\chi + a_1 \cos(\chi + \pi/3)} \qquad (3\text{-}8)$$

Parameters a_1 and τ_{cr}^* are then obtained by the least squares fitting of CRSS vs. χ dependence. It can be seen in Fig. 3-7 that Eq. 3-5 and Eq. 3-8 reproduce very closely the atomistic data for all values of χ.

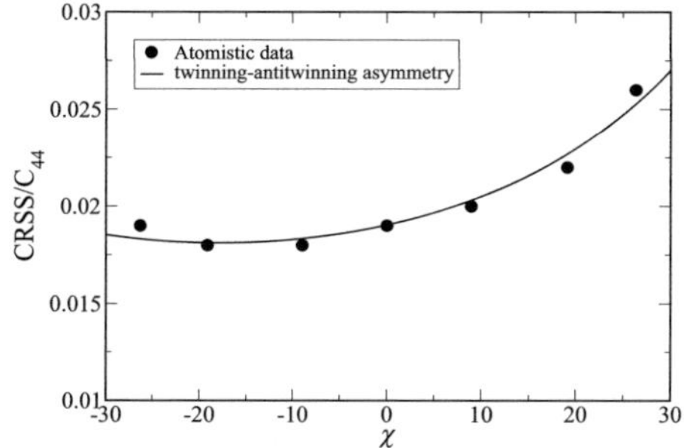

Figure 3-7. Fitting to the twinning-antitwinning asymmetry (curve) and the atomistically calculated CRSS for pure shear stress parallel to the slip direction (circles).

In the second step, keeping a_1 and τ_{cr}^* fixed, the parameters a_2 and a_3 are determined by fitting the CRSS vs. τ dependence found in the atomistic calculations using the combined stress tensor with shear stresses perpendicular and parallel to the slip direction (cf. Fig. 3-5 in Chapter 3.1.3). In this case,

$$\tau^{(\bar{1}01)} = \tau\sin(2\chi) \qquad (3\text{-}9)$$

and

$$\tau^{(0\bar{1}1)} = \tau\cos(2\chi + \pi/6) \qquad (3\text{-}10)$$

For a given angle χ and shear stress τ, the CRSS can be determined by Eq. 3-3:

$$CRSS(\chi,\tau) = \frac{\tau_{cr}^* - \tau[a_2 \sin(2\chi) + a_3 \cos(2\chi + \pi/6)]}{\cos\chi + a_1 \cos(\chi + \pi/3)} \qquad (3\text{-}11)$$

The coefficients a_2 and a_3 are again determined by the least squares fitting of this relation to the CRSS vs. τ dependencies calculated with various χ. In [109] the yield criterion for molybdenum and tungsten was fitted to atomistic data for $|\tau/C_{44}| \leq 0.02$ and only three orientations of the MRSSP, namely $\chi = 0°$ and $\chi = \pm9°$. The reason for using this limited range of atomistic data was two-fold. First, the CRSS vs. τ dependences for Mo and W were linear only for $|\tau/C_{44}| \leq 0.02$. Second, the fitting data used were limited to those for which the dislocation glided on the $(\bar{1}01)$ plane. The limitation of yield criterion fitted to such reduced data set is that it can hardly produce correctly the CRSS for the anomalous slip, for which the angle between the MRSSP and real slip plane is larger than $30°$. For example, in the atomistic simulations in Chapter 3.1.3, when the shear stress, τ, perpendicular to the slip direction is smaller than $-0.02C_{44}$, the dislocation glides on the $(0\bar{1}1)$ plane instead of the most highly stressed $(\bar{1}01)$ plane. In this case the angle between the MRSSP and the glide plane is larger than $30°$. It therefore requires the yield criterion to cover not only the glide on the primary $(\bar{1}01)$ plane but also the anomalous slip on the other $\{110\}$ planes in the [111] zone. Since the CRSS for iron with all MRSSP orientations in our atomistic simulations present a good linear dependence on τ, even when $\tau \leq -0.02C_{44}$ where the slip plane changes to the $(0\bar{1}1)$ plane, it allows us to fit to all the atomistic data with $-0.04C_{44} \leq \tau \leq 0.04C_{44}$. Moreover, our fitting database includes five MRSSP orientations

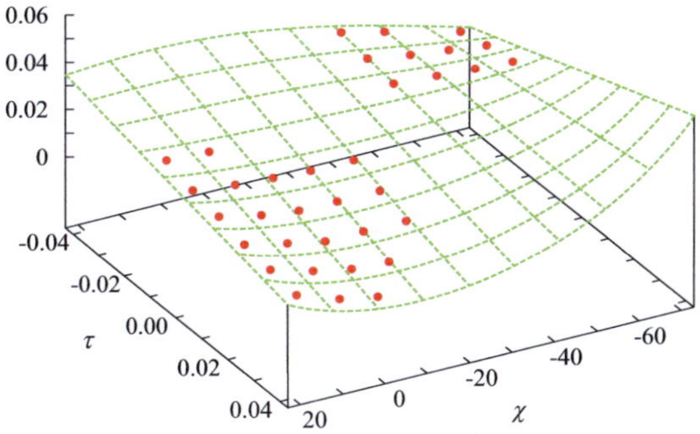

Figure 3-8. Fitting of the full yield criterion in Eq. 3-3 (dashed surface) to the atomistic data (dots).

instead of three, namely $\chi = 0°$, $\chi = \pm 9°$ and $\chi = \pm 19°$ (Fig. 3-8). The angle between the MRSSP and the glide plane, χ, is no longer limited by $-30° \leq \chi \leq 30°$ but extends to $-90° \leq \chi \leq 90°$. For anomalous slips in Fig. 3-5, the MRSSP orientations are determined as:

$$\chi^{(0\bar{1}1)} = -\frac{\pi}{3} - \chi^{(\bar{1}01)} \tag{3-12}$$

where $\chi^{(\bar{1}01)}$ is the angle between the MRSSP and the $(\bar{1}01)$ plane in the [111] zone. The positive direction of χ used in Eq. 3-12 is defined according to the twinning-antitwinning asymmetry, e.g., the positive shear stress parallel to the slip direction in the zone bounded by the $(0\bar{1}1)$ and $(\bar{1}01)$ planes are in twinning sense where χ should be negative. The coefficients

Table 3-2. Coefficients in the yield criterion (Eq. 3-3) for iron determined by fitting to the atomistic results at 0 K.

a_1	a_2	a_3	τ_{cr}^{*}/C_{44}
0.4577	0.1454	0.5645	0.0234

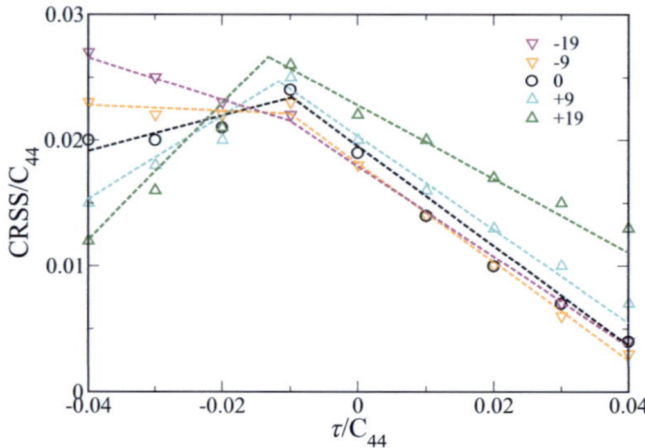

Figure 3-9. Comparison between the predictions from the yield criterion (dashed lines) and results from the atomistic calculations (points),
for $\chi = 0°$, $\chi = \pm 9°$ and $\chi = \pm 19°$.

a_1, a_2, and a_3, as well as τ_{cr}^* entering the yield criterion Eq. 3-3 for iron,

which were determined as described above, are listed in Table. 3-2.

The generalized yield criterion presents a convenient and efficient way to

predict the yielding of the iron single crystal at 0K. However, since the

accuracy of the yield criterion in reproducing the atomistic results is criti-

cal for all subsequent calculations that are based upon it, it is necessary to

perform an extensive validation tests. In Fig. 3-9, the predicted yielding

surfaces for the commencement of the motion of the $a_0/2[111]$ screw dis-

location are plotted by dashed lines for five MRSSP orientations, namely

$\chi = 0°$, $\chi = \pm 9°$ and $\chi = \pm 19°$, together with atomistic results shown by

symbols. Each point on the yielding surface corresponds to a pair of criti-

cal stresses σ_α and τ_α in the MRSSP-τ graph at which the dislocation

starts to glide. At the crossing points of the dashed lines, the slip systems

change between $\alpha = 2$ (right) and $\alpha = 13$ (left). It can be seen that the

agreement between the predictions of the yield criterion and the atomistic results is very good, not only for the magnitudes of the CRSS but also for the slip planes. This implies that both the normal slip on the $(\bar{1}01)$ primary glide plane and the anomalous slips on the inclined $\{110\}$ planes can be reproduced reliably by our yield criterion.

3.2.3 Yielding polygons for single crystal

The yield criterion formulated in the last section can be used to obtain the CRSS vs. τ dependencies for real single crystals of iron. Since any of the 24 $\{110\}<111>$ systems could be activated, the yielding can be regarded as the first commencement of the dislocation motion on the most favorable slip system. In order to identify this slip system and its CRSS value, the MRSSP for the eight distinct slip directions need to be determined first using the following formula

$$\mathbf{n}_\alpha^{\mathrm{MRSSP}} = \mathbf{l}_\alpha \times [(\mathbf{b}_\alpha \cdot \Sigma) \times \mathbf{l}_\alpha] \tag{3-13}$$

where $\mathbf{n}_\alpha^{\mathrm{MRSSP}}$, $\mathbf{l}_\alpha$, $\mathbf{b}_\alpha$, and Σ are the directions of the MRSSP normal, the dislocation line, the Burgers vector and the externally applied stress tensor respectively. The function in the square bracket is the Peach-Koehler force [52, 175] which drives the dislocation to move. The externally applied stress tensor needs to be transformed into the right-handed MRSSP$_\alpha$ coordinate system with the z-axis parallel to the corresponding $<111>$ direction ($\mathbf{l}_\alpha$ or $\mathbf{b}_\alpha$), the y-axis parallel to the direction of the MRSSP ($\mathbf{n}_\alpha^{\mathrm{MRSSP}}$), and the x-axis in the MRSSP$_\alpha$ (or parallel to the direction of the Peach-Koehler force). In general, all components of the transformed stress tensor can be nonzero. However, as already proved by the atomistic studies, the only stress components affecting the glide of the screw dislocations are the resolved shear stresses σ_α and τ_α, parallel and perpendicular to the slip direction of the reference system α, respectively. Hence, the full transformed stress tensor can be reduced to a form [109]:

$$\Sigma^\alpha(\chi_\alpha) = \begin{bmatrix} -\tau_\alpha & 0 & 0 \\ 0 & \tau_\alpha & \sigma_\alpha \\ 0 & \sigma_\alpha & 0 \end{bmatrix} \tag{3-14}$$

which contains only the values of σ_α and τ_α that enter the yield criterion. As explained above, only positive shear stresses parallel to the slip directions are considered, thus all reference systems for which σ_α are negative are excluded. This means that only four slip directions remain in the subsequent analysis. As the applied loading Σ increases from zero to the critical value (at which the crystal yields), the shear stresses σ_α and τ_α develop accordingly. The stress tensor $\Sigma^\alpha(\chi_\alpha)$ then defines a unique dependence of σ_α and τ_α, which is called the loading path characterized by $\eta_\alpha = \tau_\alpha / \sigma_\alpha$. Since for a given χ_α the CRSS_α depends only on the value of τ_α and not on the history how the corresponding combination of σ_α and τ_α was achieved, the yielding criterion can be expressed using the ratio η_α as:

$$CRSS_\alpha = \frac{\tau_{cr}^*}{\cos\chi_\alpha + a_1\cos(\chi_\alpha + \pi/3) + \eta_\alpha[a_2\sin(2\chi_\alpha) + a_3\cos(2\chi_\alpha + \pi/6)]}$$

(3-15)

The angle χ_α is the angle the MRSSP makes with the corresponding $\{110\}$ reference plane for a glide system α. Note that the range of this angle is no longer limited by $-30° \leq \chi \leq 30°$ but $-90° \leq \chi \leq 90°$ to accommodate possible anomalous slips. The sign of χ_α is defined according to the sense of the shearing, i.e., $\chi_\alpha < 0$ corresponds to shearing in the twinning sense while $\chi_\alpha > 0$ corresponds to shearing in the antitwinning sense. This definition conforms to that used in the atomistic studies in the last section where $\alpha = 2$. Then a reference system α is considered to become activated when σ_α increases to the $CRSS_\alpha$. The system with the lowest critical loading is the active slip system on which the plastic deformation commences first.

To develop a complete description of the yielding of a single crystal, the procedure described above has to be repeated for all possible loading paths, i.e. $-\infty < \eta < +\infty$. For each loading path four reference systems, each associated with a distinct slip direction, are evaluated, and for every slip system the yielding point (a pair of critical stresses σ_α and τ_α in the MRSSP-τ graph) is determined. In order to obtain a clear view of the yielding, it is convenient to project the yielding points for systems other than $(\bar{1}01)[111]$ into the CRSS-τ graph for the $(\bar{1}01)[111]$ reference system. The point on the straight loading path η starting from the origin of the CRSS-τ graph that is closest to the origin marks the stress that causes the primary {110}<111> system to become activated, or, equivalently, the stress at which the single crystal starts to deform plastically. The lines connecting the points of minimum CRSS along all loading paths then compose the yield polygon, which is the projection of the yield surface on the MRSSP-τ graph for a certain χ_α in system α. This analysis can be carried out for each of the 24 slip systems in any orientation of MRSSP.

In Fig. 3-10, the critical points marking the onset of activation, predicted following the procedure described above, are plotted for $\chi = 0°$, $\chi = \pm 9°$ and $\chi = \pm 19°$ in the slip system $(\bar{1}01)[111]$ ($\alpha = 2$). The colors are used to distinguish between different slip systems. The projections of the yield surface on these MRSSP-τ graphs are the inner polygons surrounded by the solid lines.

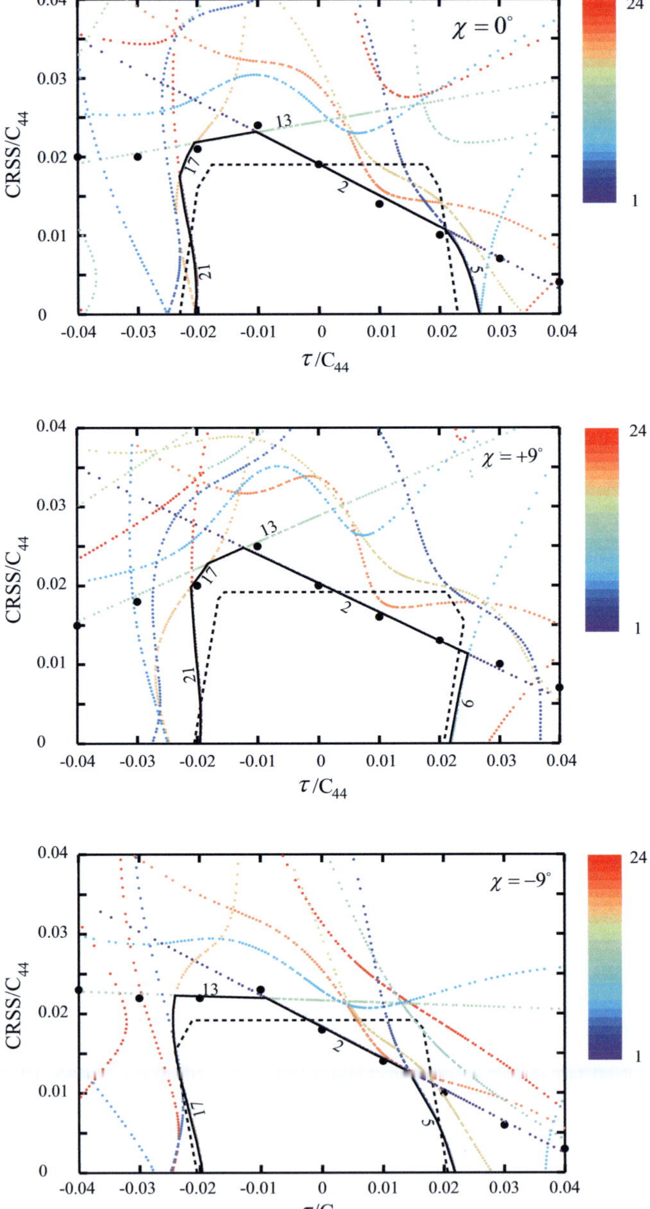

χ = 0°
CRSS/C_44
τ/C_44
24
1
13
17
2
21
5
χ = +9°
CRSS/C_44
τ/C_44
24
1
13
17
2
21
9
χ = −9°
CRSS/C_44
τ/C_44
24
1
13
2
17
5

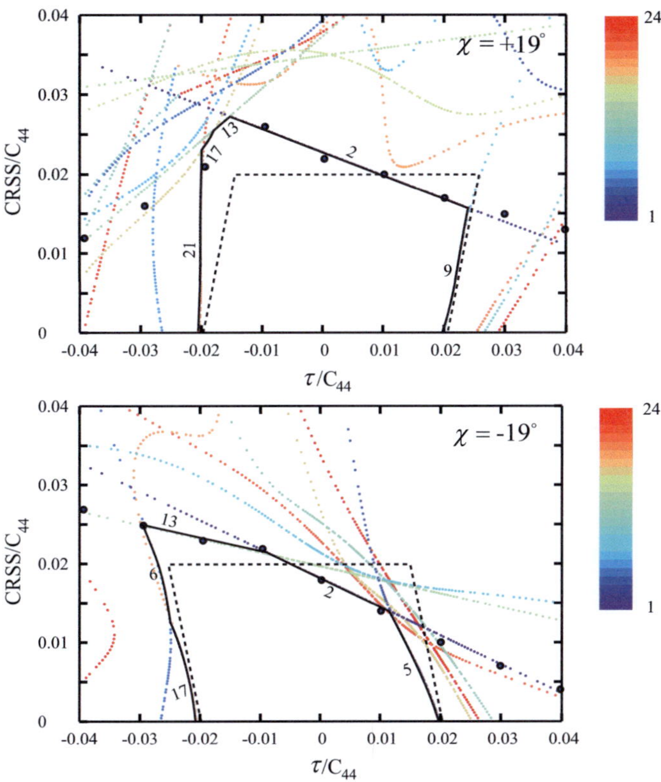

Figure 3-10. The yield surfaces projected on the $(\bar{1}01)[111]$ slip system for $\chi = 0°$, $\chi = \pm 9°$ and $\chi = \pm 19°$, for all slip systems distinguished by different colors (legend bar). The inner solid polygons indicate the active slip systems predicted by the yield criterion while the dashed one is from the Schmid law. The black dots are the atomistic results in Secction 3.1.3.

We see from these plots how the glide plane changes for different loading paths, η. If the magnitude of the shear stress perpendicular to the slip direction is small, roughly $-0.01C_{44} \leq \tau \leq 0.02C_{44}$, the primary slip system coincides with the most highly stressed $(\bar{1}01)[111]$ system. However, as $|\tau|$ becomes larger, other $\{110\}\langle 111\rangle$ systems become dominant. Since the values of τ at which the plastic deformation of real crystals takes place are bounded by the yield polygon, $|\tau|$ can never be larger than $\sim 0.03 C_{44}$. If the loading path falls within a close vicinity of the corners of the inner polygon, more slip systems become activated simultaneously and a multiple slip occurs. For example, in the atomistic simulations in Section 3.1.3 [see Fig. 3-5(a)] two slip systems, namely $\alpha = 2$ and 13, are observed to be activated simultaneously under loading with $\chi = 0$ and $\tau \sim -0.01 C_{44}$. This feature is correctly predicted by our yield criterion in Fig. 3-10(a) where the purple dot line ($\alpha = 2$) intersects with the green dot line ($\alpha = 13$) at one of the corners of the yield polygon where $\tau \sim -0.01 C_{44}$.

For illustration, it also shows by dashed lines how the projection of the yield surface looks if the effective yield criterion reduces to the Schmid law. In this case, the CRSS for the most highly stressed $(\bar{1}01)[111]$ system is independent of τ. At larger τ, the yield polygon is bounded by the inclined critical lines that correspond to different reference systems other than $(\bar{1}01)[111]$. These critical lines are inclined because of the projection, which does not mean that the CRSS is a function of τ at large τ. Since the $(\bar{1}01)$ plane is a mirror plane in bcc crystals, the Schmid-law yield surface projected in the CRSS-τ graph for $\chi = 0$ is completely symmetrical with respect to $\tau = 0$. In comparison, the yield polygons for $\chi = \pm 9^\circ$ are not symmetrical with respect to $\tau = 0$ since their MRSSP are no longer coincident with the $(\bar{1}01)$ mirror plane. However, since the Schmid factor is an odd function of the MRSSP orientation, the yield polygons predicted using only the Schmid law for $\chi = \pm 9^\circ$ are mirror images of each other. All these symmetries are broken in iron if the non-Schmid stresses are considered. As shown in Fig. 3-10 as a consequence of the twinning-antitwinning asymmetry and the strong effect of the shear stresses perpendicular to the slip direction, the real yield behavior in single crystal iron is much more complex than that predicted from the Schmid law.

Up to now following the atomistic simulations on the motion of a single $a_0/2[111]$ screw dislocation in bcc iron presented in section 3.1, we studied the macroscopic yielding of single crystals containing $a_0/2<111>$ screw dislocations with all possible Burgers vectors based upon the yield criterion, which can closely reproduce the atomistic results. The constructed yield criterion shows its ability in predicting the commencement of the motion of the $a_0/2<111>$ screw dislocations under external loadings at 0 K. One should note the yield criterion did not include any temperature effects. However it has been proved experimentally that the temperature dependence of the yield stress is very important in plastic deformations of bcc metals. Thus in the next section a link between the achieved results at 0 K and the thermally activated motion of the $a_0/2<111>$ screw dislocations will be developed.

3.3   Thermally activated motion of screw dislocation

According to the atomistic studies, owing to the non-planar core structure of the $a_0/2<111>$ screw dislocations, the lattice resistance is very high at 0 K compared with that of dislocations in face-centered cubic metals [26,

44, 53, 56, 81, 82, 170, 176, 177]. This indicates a strong Peierls barrier between two neighboring stable sites of the periodic lattice that the dislocation has to overcome [52, 110-113]. The Peierls stress determined from the static atomistic calculations presents the limiting value at zero temperature. However, it is observed experimentally that the yield stress decreases with increasing temperature. In the following, we will concentrate on the motion of the screw dislocations at finite temperatures between 0 K and the critical temperature T_k.

As mentioned in Chapter 1, in this region the Peierls barrier for the straight dislocation can be surpassed with the aid of thermal activation via nucleation of kink-pairs, which subsequently migrate relatively easier along the dislocation line [52, 120, 121], so that a part of the energy needed to activate the dislocation is supplied by thermal fluctuations. This part of energy used in the formation of the kink-pairs is called activation enthalpy, which is a function of the applied stress according to the transition state theory of thermally activated processes [122-124]. One way to obtain the activation enthalpy in terms of stress is to investigate the activation path between two neighboring stable positions of the screw dislocation using the NEB method (for example [131-134]). An alternative approach is to study the glide of the $a_0/2<111>$ screw dislocations at finite temperatures by means of molecular dynamic simulations [91, 92, 133, 135, 136]. However, as discussed already in Chapter 1, both methods are problematic for loadings with arbitrary orientation, when considering the effects of non-Schmid stresses.

Instead in this chapter, a phenomenological description of the Peierls potential for the $a_0/2<111>$ screw dislocation in iron will be developed following a recent work of Gröger and Vitek [138]. The main advantage of the constructed Peierls potential is that it can reflect the dependence of the Peierls stress on the MRSSP orientation, χ, and the shear stresses perpendicular to the Burgers vector [138]. The thermally activated dislocation motion can then be treated using the line tension (LT) model at low temperatures and elastic interaction (EI) model at high temperatures [121, 122]. In the following, it first introduces the construction of the Peierls potential based on the yield criterion developed in Chapter 3.2. Using the constructed Peierls barrier, the activation enthalpy for the formation of kink-pairs can be determined as a function of the applied stress tensor, and the corresponding temperature and strain rate dependence of the yield stress can be evaluated using the Arrhenius equation (Eq. 1-1).

3.3.1 Construction of the Peierls potential and the Peierls barrier

In the high stress regime (at low temperatures), according to the classical Peierls potential model, the dislocation is first shifted as a straight line along the Peierls potential, which is a one-dimensional periodic function of the reaction coordinates [Fig. 1-2(b)]. With the aid of thermal fluctuations, segments of the dislocation vibrate and bow out to various intermediate configurations. This applies to dislocations in fcc metals which have planar core structures and thereafter specified glide planes. However, owing to the non-planar core structure, the $a_0/2\langle 111\rangle$ screw dislocations in bcc metals do not have unique slip planes and the Peierls barrier is also a function of the core transformation. Hence, in the current work it follows the suggestions in [137], where the Peierls potential, $V(x, y)$, is regarded as a function of two variables, x and y, which represent the position of the intersection of the dislocation line with the {111} plane perpendicular to

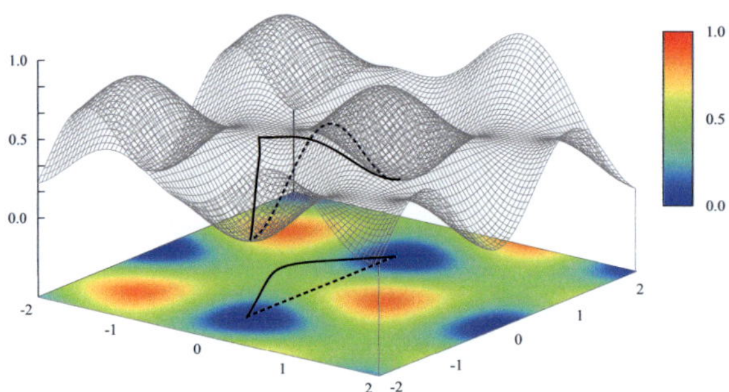

Figure 3-11. Contour plot of the mapping function $m(x, y)$ determined by Eq. 3-16. The straight dashed line is the initial path between two neighboring stable sites and the solid curve is the minimum energy path of the transition. Color maps the height of the potential.

the corresponding <111> slip direction. The transition of the screw dislocation between two stable sites at 0K is then regarded as a motion along the MEP, described by a coordinate ξ of the Peierls barrier $V(\xi)$. In this representation, the Peierls barrier is dependent on the Peierls potential, which will be a function of the applied stress tensor with both shear stress components parallel and perpendicular to the slip direction.

Following the first implementation of the Peierls potential by Edagawa et al. [137] and then developed by Gröger et al. [138], to capture the three-fold rotation symmetry associated with <111> directions, the Peierls potential is based on the product of three sinusoidal functions. For the screw dislocation along the [111] direction this so-called m-function can be expressed as:

$$m(x,y) = \frac{1}{2} + \frac{4}{3\sqrt{3}}\sin\frac{\pi}{3a}(2y\sqrt{3}+a)\sin\frac{\pi}{a}(\frac{y}{\sqrt{3}}-x-\frac{a}{3})\sin\frac{\pi}{a}(\frac{y}{\sqrt{3}}+x+\frac{2a}{3})$$

(3-16)

where (x, y) are the coordinates along the $[\overline{1}2\overline{1}]$ and $[\overline{1}01]$ directions, respectively. The m-function is depicted as a contour plot in Fig. 3-11, where the blue shading corresponds to minima and the red shading to maxima. It is threefold symmetric with extreme $0 \leq m(x, y) \leq 1$. The minima and maxima of $m(x, y)$ form a triangular lattice with the lattice parameter $a = a_0\sqrt{2/3}$, where a_0 is the lattice constant of the bcc lattice. The Peierls barrier, $V(\xi)$, is regarded as energetic maximum of the minimum energy path between two neighboring stable sites with lowest energy in the two dimensional Peierls potential field $V(x, y)$.

For a given potential $V(x, y)$, the coordination path ξ can be determined using the NEB method. The link between the Peierls barrier and the Peierls stress, σ_p, is:

$$\sigma_p b = \max[\frac{\mathrm{d}V(\xi)}{\mathrm{d}\xi}]$$

(3-17)

which is the fundamental relationship that allows us to construct the Peierls potential as a function of the applied stress tensor, based on the results of the atomistic studies.

The development of the Peierls potential includes the determination of the height of the potential without external loading, the dependence of the potential on the shear stress parallel to the slip direction and the dependence on the shear stress perpendicular to the slip direction. They are achieved in steps in a self-consistent manner using NEB method [138].

(1)   Height of the Peierls potential

The height of the Peierls potential is simply set as:

$$V(x,y) = V_0 m(x,y) \qquad\qquad (3\text{-}18)$$

where V_0 is the maximum height of the potential. The prefactor V_0 is determined by the following self-consistent procedure. First, a trial value of V_0 is chosen and the NEB method is used to determine the minimum energy path, ξ, between the adjacent minima along the $(\bar{1}01)$ glide plane. Using the Peierls barrier $V(\xi)$ obtained in this way, $\max[dV(\xi)/d\xi]$ is evaluated and compared with σ_{p}, which is the CRSS of the loading with $\chi = 0$ for pure shear stress parallel to the $[111]$ direction in the $(\bar{1}01)$ plane, determined by either atomistic simulation or the yield criterion. We then adjust V_0 and repeat the whole process until the difference between $\max[dV(\xi)/d\xi]$ and σ_p becomes less than 10^{-4} eV/Å^2. The height of the Peierls potential in Eq. 3-18 for iron determined by this approach yields $V_0 = 0.05195$ eV.

(2)   Effect of the shear stress parallel to the slip direction

If the Peierls potential was independent on the applied stress tensor, the orientation dependence of the CRSS would follow exactly the Schmid law as CRSS $\sim 1/\cos\chi$. However as shown in Fig. 3-3, this is not true for Fe and other bcc metals. Providing that only the shear stress parallel to the slip direction is applied, the CRSS varies with the orientation of the MRSSP in such a way that it is higher for the antitwinning shear ($\chi > 0$) and lower for the twinning shear ($\chi < 0$), relative to the value for $\chi = 0$ when the MRSSP coincides with the $(\bar{1}01)$ plane. The orientation dependence of the CRSS implies that the activation energy barrier for the motion of the dislocation is higher when $\chi > 0$ and lower when $\chi < 0$. This effect can be implemented into Peierls potential in the following way.

With the fixed V_0 achieved in the previous step, the dependence of the Peierls potential on the shear stress parallel to the slip direction can be expresses as:

$$V(x,y) = m(x,y)[V_0 + V_\sigma(\chi,\theta)] \tag{3-19}$$

where the Peierls barrier now also varies with the orientation of the MRSSP and the magnitude of the shear stress σ via an angularly dependent function:

$$V_\sigma(\chi,\theta) = K_\sigma(\chi)\sigma b^2 \cos\theta \tag{3-20}$$

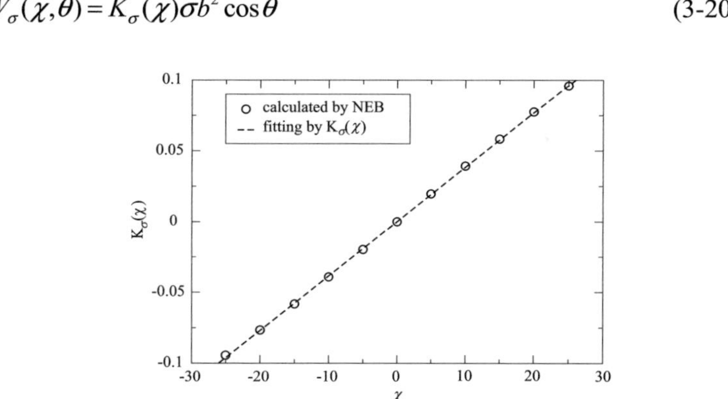

Figure 3-12. Fitting the dependence of K_σ to the MRSSP angle χ.

where θ is the angle between the *x*-axis and the line connecting the origin and the point (x, y). The function $K_\sigma(\chi)$ is determined in a similar self-consistent way as the height V_0 : for a given χ, it starts with an initial guess of K_σ to obtain a trial Peierls potential using Eqs. 3-19 and 3-20. The

NEB method is then used to find the minimum energy path, ξ, between two adjacent potential minima on the $(\bar{1}01)$ plane. Then $\max[dV(\xi)/d\xi]$ is evaluated for the Peierls barrier $V(\xi)$ obtained in this way and compared with $\sigma_p b$ for which

$$\sigma_p = CRSS(\chi)\cos\chi \qquad (3\text{-}21)$$

and

$$CRSS(\chi) = \frac{\tau_{cr}^*}{\cos\chi + a_1\cos(\chi + \pi/3)} \qquad (3\text{-}22)$$

which is the yield criterion for loadings with only pure shear stress parallel to the slip direction. K_σ is then adjusted and the whole process is repeated until the Peierls stress, σ_p, is reproduced with the precision of 10^{-4} eV/Å^2 compared to the value achieved by the yield criterion.

Fig. 3-12 shows the value of K_σ as a function of a set of orientations of the MRSSP. Apparently, the value of K_σ depends linearly on the MRSSP angle.

Thus, $K_\sigma(\chi)$ can be well approximated by a linear function:

$$K_\sigma(\chi) = k\chi \qquad (3\text{-}23)$$

When $\chi = 0$, no non-glide stresses are present and $K_\sigma(\chi)$ becomes zero. In this case, the dislocation glide is governed by the Schmid law, and the Peierls potential (Eq. 3-19) reduces to that given by Eq. 3-18. For positive χ, i.e. shearing in the antitwinning sense, $V_\sigma(\chi,\theta)$ is positive, and both the Peierls barrier and the Peierls stress for the $(\bar{1}01)$ slip increase relative to χ

= 0. In contrast, for negative χ, i.e. twinning shear, $V_\sigma(\chi,\theta)$ is negative and both the Peierls barrier and the Peierls stress decrease compared to $\chi =$ 0. Therefore, Eq. 3-19 represents the Peierls potential that reproduces the twinning-antitwinning asymmetry of glide for loading by the shear stress parallel to the slip direction.

Besides, the term $V_\sigma(\chi,\theta)$ in Eq. 3-20 is also capable to reflect the symmetry operation of the twinning-antitwinning effect. For example, upon reversing the sense of shearing, Eq. 3-20 becomes

$$V_{-\sigma}(\chi,\theta) = K_\sigma(\chi)(-\sigma)b^2$$
$$= K_\sigma(-\chi)\sigma b^2 \qquad (3\text{-}24)$$
$$= V_\sigma(-\chi,\theta)$$

This implies that the reversal of the sense of shearing is identical to keeping the stress and reversing the sign of the angle χ. This is why in Fig. 3-3 (Chapter 3.1.2), when comparing the CRSS of compression with that of tension for the same loading orientation, the angle of MRSSP, χ, has to be reversed.

**(3)   Effect of the shear stress perpendicular
to the slip direction**

In order to incorporate the effect of the shear stress, τ, perpendicular to the slip direction, $V(x, y)$ is supplemented by a third term $V_\tau(\chi,\theta)$, which represents the distortion of the Peierls potential by τ. The Peierls potential that comprises the effects of both the shear stress parallel and the shear stress perpendicular to the slip direction is then:

$$V(x,y) = [V_0 + V_\sigma(\chi,\theta) + V_\tau(\chi,\theta)]m(x,y) \qquad (3\text{-}25)$$

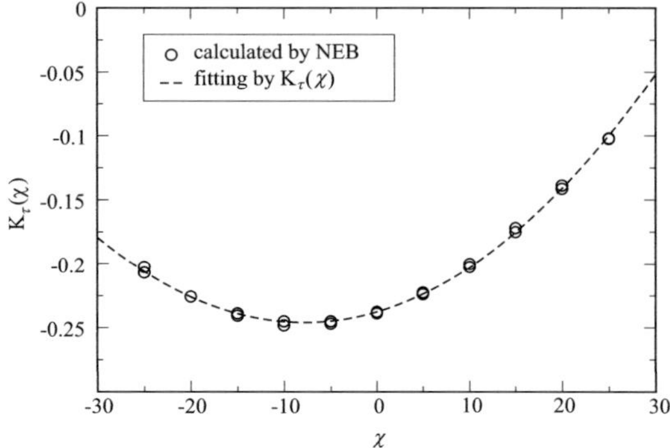

Figure 3-13. Fitting the dependence of $K_\tau(\chi)$ to the MRSSP angle, χ, with two values of τ.

for which

$$V_\tau(\chi,\theta) = K_\tau(\chi)\tau b^2 \cos(2\theta + \pi/3) \qquad (3\text{-}26)$$

$K_\tau(\chi)$ can be determined for a given χ in a similar way as $K_\sigma(\chi)$. One can start again with an initial guess of $K_\tau(\chi)$ and determine a trial Peierls potential by Eqs. 3-25 and 3-26. The values of V_0 and $K_\sigma(\chi)$ entering Eq. 3-25 are those determined as described previously. The minimum energy path, ξ, between two adjacent potential minima on the $(\overline{1}01)$ glide plane is then determined using the NEB method. For this path, $\max[dV(\xi)/d\xi]$ is evaluated and compared again with $\sigma_p b$, which can be determined using Eq. 3-21 and

$$CRSS(\chi,\tau) = \frac{\tau_{cr}^* - \tau[a_2 \sin(2\chi) + a_3 \cos(2\chi + \pi/6)]}{\cos\chi + a_1 \cos(\chi + \pi/3)} \qquad (3\text{-}27)$$

This equation is the complete yield criterion for loadings with pure shear stresses both parallel and perpendicular to the slip direction. $K_\tau(\chi)$ is adjusted accordingly and the whole process is then repeated until the value of $K_\tau(\chi)$, for which the Peierls stress σ_p is reproduced with the precision of 10^{-4} eV/Å^2 comparing to the value attained by the yield criterion, is obtained.

In order to keep the calculation of $K_\tau(\chi)$ simple, only two different shear stresses perpendicular to the slip direction, namely $\tau = \pm 0.01 C_{44}$, are considered. The dependence of $K_\tau(\chi)$ on χ is plotted in Fig. 3-13.

For iron, $K_\tau(\chi)$ can be closely approximated by a quadratic polynomial:

$$K_\tau(\chi) = C_0 + C_1\chi + C_2\chi^2 \tag{3-28}$$

This variation of the Peierls potential reflects the transformation of the dislocation core and consequently changes of the glide plane, as will be discussed in more detail in Chapter 4.1.

Fig. 3-14 shows contour plots of the final Peierls potentials (Eq. 3-25) for three different loadings with only shear stress τ perpendicular to the slip direction. In these plots the blue domains correspond to minima and red domains to maxima. The corresponding minimum energy paths between adjacent potential minima, determined by the NEB method, are superimposed as dashed curves. It is obvious that positive shear stress perpendicular to the slip direction lowers the potential barrier for the slip on both $(\bar{1}01)$ and $(\bar{1}10)$ planes. Considering the Schmid factor, the resolved shear stress on the $(\bar{1}01)$ plane is higher than that on the $(\bar{1}10)$ plane; the dislocation therefore prefers to glide on the $(\bar{1}01)$ plane. In contrast, for negative τ the

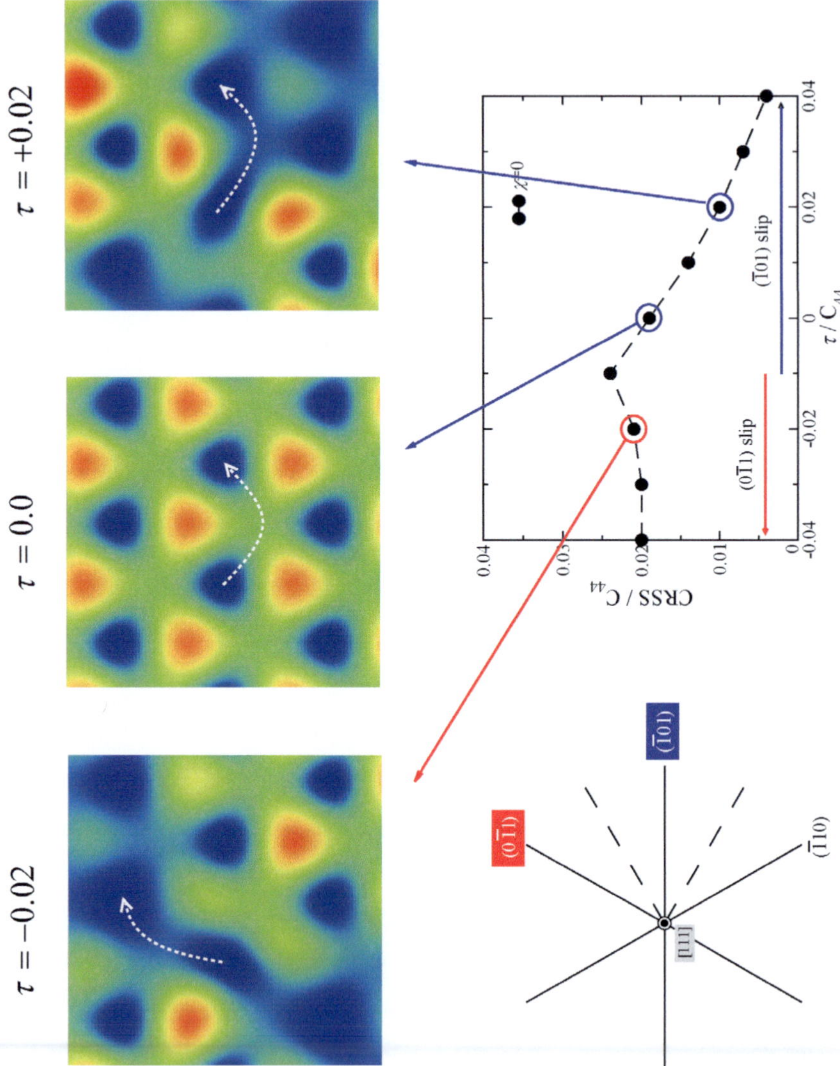

Figure 3-14. Contour plots of the Peierls potential for three different applied loadings with only shear stress τ perpendicular to the slip direction, in which the blue domains correspond to minima and red domains to maxima. The active reaction paths with lowest energy are drawn as dashed curves. For comparison, the corresponding atomistic data and their slip orientations are given in the lower panel.

Peierls barrier for the $(\bar{1}01)$ and $(\bar{1}10)$ slip are higher than that for the $(0\bar{1}1)$ plane, so the glide on the latter is more likely. For $\tau = 0$, although the Peierls barriers for the three glide planes are the same, since the Schmid factor on the $(\bar{1}01)$ plane is two times higher than that on the other two planes, the dislocation prefers to glide on the $(\bar{1}01)$ plane. All these conclusions agree perfectly with the findings of atomistic simulations, which demonstrate that the predictions based on the Peierls potential are consistent with the results of atomistic calculations at 0K.

It should be noted that the Peierls barrier $V(\xi)$ obtained from the *m*-function has a sharp maximum due to its sinusoidal character. However, it has been shown in [117, 178-182] that a better agreement between calculated temperature dependence of the yield stress and the experimental data is obtained if the Peierls barrier is flat, i.e. the MEP has a flat plateau instead of the sharp maximum. In Chapter 4, it will show in detail how the constructed Peierls potential with flat top, which provids better agreements between our predictions and experimental results, is developed by using a flatting operator $\hat{f}$.

3.3.2 Stress dependence of the activation enthalpy

Two models for thermally activated dislocation motion at both high and low temperatures were introduced in Chapter 1. At high temperatures, fully developed kink pairs are formed by thermal fluctuations. Once a critical configuration is reached, the kinks propagate and consequently the screw dislocation moves. The activation enthalpy $H(\sigma)$ required to reach this critical point can be determined in terms of the elastic Eshelby attraction between these two fully developed kinks and the repulsive interaction between them produced by the external loading. In this case the activation enthalpy is only dependent on the shear stress component σ^*, which is the projection of the applied stress on the {110} glide plane for a given <111>{110} slip system and parallel to the slip direction. The shear stress perpendicular to the slip direction does not affect the activation enthalpy, indicating it is independent on the shape of the Peierls barrier. To determine the activation enthalpy, H_{kp}, at high temperatures using Eq. 1-4, one needs the energy of an isolated kink H_k. This energy can be either calculated atomistically, as was done by Duesbery [125, 126] or estimated from experiments. For iron the value of $2H_k$ determined experimentally equals to 0.927 eV by Brunner [118]. Theoretical calculations predict values between 0.6 and 1.1 eV [92, 183].

At low temperatures, it is assumed that the straight dislocation is pushed up from its equilibrium position in the Peierls potential valley by high applied stresses, and then bows out by thermal fluctuations. When the bow-out reaches a critical configuration it continues to expand as a fully developed kink-pair and the dislocation moves forward. The activation enthalpy can then be determined using Eq. 1-9 by integrating the Peierls barrier energy in terms of the coordinate ξ, and then subtracting the work done by the shear stress σ^*, which is the projection of the applied stress on the glide plane and parallel to the slip direction. In contrast to the high temperature model, for the low temperature model the Peierls potential is a function of the MRSSP orientation and the stress components both parallel and perpendicular to the slip direction (Eq. 3-25). Consequently, for a given <111>{110} slip system the activation enthalpy is a function of the full applied stress tensor, unlike in the high-temperature regime, in which it is dependent on the shear stress σ^* only. To determine the activation enthalpy H_b using Eq. 3-25 at low temperatures, one needs the value of the line tension E. Since there is no data from either atomistic simulations or experiments, a theoretical value is used, i.e. $E \sim \mu b^2/4$ [138], where b is the magnitude of the Burgers vector and μ is the shear modulus for the <111>{110} slip system studied which can be determined as $(C_{11} - C_{12} + C_{44})/3$.

Up to now all parameters required to determine the activation enthalpy for the thermally activated motion of the $a_0/2$<111>{110} screw dislocations are achieved and summarized in Table 3-3. The Peierls potential (Eq. 3-25) in conjunction with the line-tension model (Dorn-Rajnak expression, Eq. 1-9) at low temperatures, and the elastic-interaction model (Eq. 1-4) at high temperatures, can be used to predict the dependence of the activation enthalpy on the applied stress. The transition between the low and high temperature models occurs at the stress where the activation enthalpies coincide, i.e. $H_{kp} = H_b$.

Table 3-3. Parameters for the Peierls potential in Eq. 3-25.

V_0	k	C_0	C_1	C_2
0.0520	0.2202	-0.2376	0.1216	0.4447

Fig. 3-15 shows an example of the dependence of the activation enthalpy on the shear stress σ^* for the $(\overline{1}01)[111]$ slip system ($\alpha = 2$) loaded in tension along the $[\overline{1}49]$ direction. With this loading the corresponding MRSSP is the $(\overline{1}01)$ plane for which $\chi = 0$. The ratio of the two resolved shear stresses is $\eta = 0.51$. In the low stress region (red curve), the elastic interaction model applies. When the thermal component of the yield stress

σ^* is zero, the activation enthalpy equals to $2H_k \sim 0.927$ eV, and the dislocation is driven purely by the athermal stress $\bar{\sigma}$. As the applied stress increases, the activation enthalpy required to develop the kink-pairs decreases. The transition between the elastic interaction model at high temperatures and the line tension model at low temperatures, occurs at $\sigma^* \sim$ 250 MPa. In the low temperature region, the activation enthalpy continues to decrease with increasing applied stress. When the activation enthalpy decreases to zero, the corresponding σ^* is about 1800 MPa. One should note, when comparing to the experimental results, the Peierls stress of the screw dislocation at 0K, computed by atomistic simulations, is typically 3-5 times higher than the value estimated from experiments [34, 68, 92, 95, 98, 132, 184-187]. This kind of deviations exists between experiments and atomistic simulations for all bcc metals regardless the description scheme of the atomic interaction. Several explanations have been proposed [187], but so far none provided a satisfactory clarification of this problem (see more in Chapter 4). In order to compare the experimental and theoretical data, it is customary to rescale the calculated shear stresses. This scaling factor will be discussed in the next Chapter 4.3.

The enthalpy-stress dependence shown in Fig. 3-15 is only for the $a_0/2[111]$ screw dislocation on the $(\bar{1}01)[111]$ slip system, which possesses the highest effective Schmid factor and the lowest activation energy. Generally, when considering single crystal with dislocations of all possible Burgers vectors, the total plastic strain rate should be determined from the Arrhenius law using a summation:

$$\dot{\gamma} = \dot{\gamma}_0 \sum_{\alpha} \exp[-\frac{H^{\alpha}(\sigma)}{k_B T}] \qquad (3\text{-}29)$$

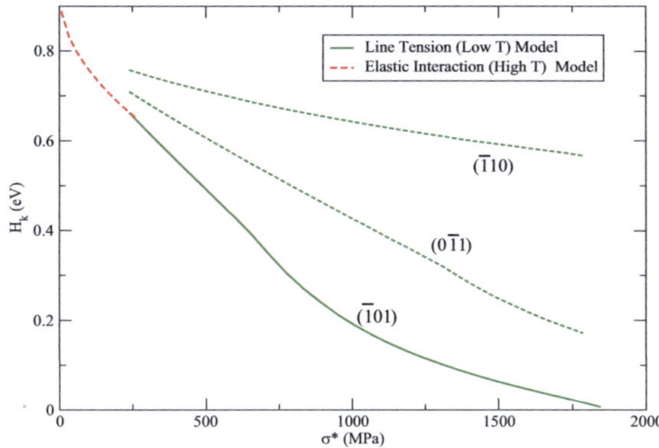

Figure 3-15. Dependence of the activation enthalpy, $H_b(\sigma)$, on the shear stress, σ^*, projected on the {110} slip planes for the $a_0/2[111]$ screw dislocation, with tensile loading in the $[\bar{1}49]$ direction. At low temperatures, the line tension model applies (green curves) and, the glide on the primary slip plane,$(\bar{1}01)$, has the lowest activation energy. At high temperatures, the elastic interaction model applies (red curve).

One should note that $\dot{\gamma}$ is not the strain rate belonging to a certain glide system but the total plastic strain rate. In general, the summation should cover all possible slip systems α that can be activated for the given loading. However, since the activation enthalpy appears in Eq. 3-29 in exponent, the contribution of most slip systems with larger $H(\sigma)$ can be safely neglected. For example, in Fig. 3-15 it shows also the activation enthalpies for the other two slip planes, i.e., $(0\bar{1}1)$ and $(\bar{1}10)$, that belong to the same [111] zone of the $a_0/2[111]$ screw dislocation. One can see that the calculated activation enthalpy for the $(\bar{1}01)[111]$ system is at any stress significantly lower than that for the other two slip systems. This means that the rate equation 3-29 is dominated by the term involving the activation enthalpy for the $(\bar{1}01)[111]$ system only. Eq. 3-29 can be therefore safely reduced to the single rate equation:

$$\dot{\gamma} = \dot{\gamma}_0 \exp\left[-\frac{H^{\alpha}(\sigma)}{k_{\mathrm{B}}T}\right] \qquad (3\text{-}30)$$

for which $\alpha = 2$.

4   Discussion

4.1   Dislocation mobility by atomistic simulations

In Chapter 3.1, the critical yield stress for the $1/2[111]$ screw dislocation was determined by means of static atomistic simulations using the BOP model. The different values of CRSS for loadings with pure shear stress parallel to the slip direction and for uniaxial loadings in tension and compression with the same MRSSP orientation clearly showed that the shear stress parallel to the slip direction is not the only stress determining the critical yield stress of the screw dislocation. The non-Schmid stresses, or more specifically the shear stresses perpendicular to the slip direction, also markedly affect the Peierls barrier.

In the previous Chapter, it describes how these atomistic results can be utilized in the formulation of phenomenological yield criteria as well as in the description of macroscopic plasticity at finite temperatures. In the present section, the underlying microscopic mechanism will be analyzed and it will show that most of the macroscopic mechanical properties can be linked to changes of dislocation core structures.

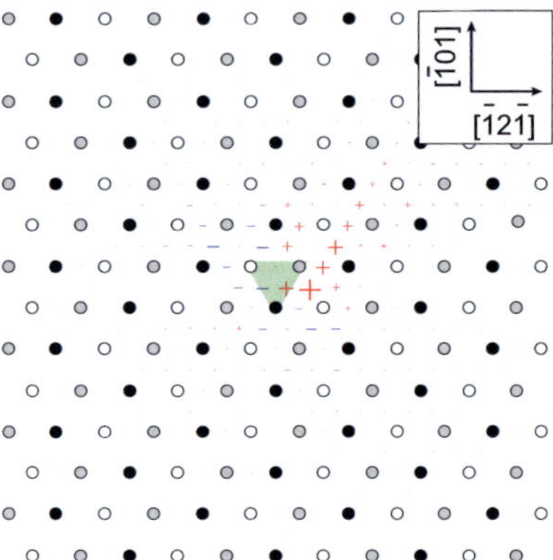

Figure 4-1. Change of the differential displacements around the core under pure shear stress equal to $0.015C_{44}$ applied on the $(\bar{1}01)$ plane. The magnitude of change is multiplied by a factor of 20. The shading is used to highlight the initial position of the dislocation center.

The differential displacement plot in Fig. 3-1 shows that under zero stress the core structure of the $a_0/2<111>$ screw dislocation in Fe spreads symmetrically on the three $\{110\}$ planes of the $[111]$ zone. However, under applied stress the dislocation core changes as a function of shear stresses parallel as well as perpendicular to the Burgers vector. Before commence of yielding, this change is purely elastic in that the structure returns into its original configuration if the stress is removed. However, once the applied shear stress σ reaches CRSS and the dislocation starts moving, the transformation is no more elastic and the gliding core remains distorted. This is in contrast to fcc metals and hexagonal crystals, where the screw dislocations (or partial dislocations) move at very low stresses without significant changes in the dislocation cores.

For pure shear stress parallel to the Burgers vector, i.e., without any non-Schmid stresses, the dislocation core gradually transforms from its symmetric non-degenerate configuration to a less symmetric form as the loading increases. It has been assumed that during the glide the 'arms' on the inclined {110} planes shorten while those on the $(\bar{1}01)$ plane become more extended rendering the core more glissile [101]. Fig. 4-1 showing the changes of the differential displacements around the core under pure shear stress on the $(\bar{1}01)$ plane equal to $0.015C_{44}$ (smaller than the Peierls stress) reveals that this assumption is not completely correct. One can see that the largest changes of the core indeed occur on the horizontal $(\bar{1}01)$ glide plane, but they are limited to the very core center only: there is a marked reduction on the left from the initial position and a large increase on the right at the final position. This corresponds to the shift of the Burgers vector from the original core position to the neighbouring stable site. The symbols in Fig. 4-1 also clearly reveal the twinning-antitwinning effect. Although the pure shear stress parallel to the slip direction is applied along the horizontal $(\bar{1}01)$ plane, which is a mirror plane, the change of the core is not symmetric in terms of the $(\bar{1}01)$ plane. The core under stress prefers to extend more in the twinning region above the $(\bar{1}01)$ plane $(-60° < \chi < 0°)$, while it contracts in the anti-twinning region below the $(\bar{1}01)$ plane $(0° < \chi < 60°)$. This also explains the twinning-antitwinning asymmetry obtained for the CRSS vs. χ.

Apart from the pure shear stress calculations, it also showed (cf. Fig. 3-3) that the CRSS for tension is always lower than that for compression in the same loading direction. The CRSS for pure shear with the same MRSSP lies in between of those for tension and compression. This is the so-called tension-compression asymmetry observed in experiments. The tension-compression asymmetry clearly indicates that the CRSS depends not only on the Schmid stresses but also on the shear stresses other than those parallel to the slip direction. More specifically, the glide of the screw dislocation is affected by the shear stress perpendicular to the slip direction. Although this stress component does not drive directly the screw dislocation to move, it changes the symmetry of the core and makes the dislocation either easier or harder to slip on different {110} planes. The change of the dislocation core in bcc molybdenum under shear stress perpendicular to the slip direction for the $\chi = 0$ orientation was examined in [101]. The authors found that the core indeed either extended or constricted on the glide plane and, as a consequence, its motion was either promoted or suppressed. In order to investigate the dependence of the CRSS on the magnitude of the shear stress τ perpendicular to the slip direction, a set of simulations of the $a_0/2[111]$ screw dislocation subjected to loadings with various combination of shear stresses both parallel and perpendicular to the slip direction, as described by Eq. 3-2, are carried out.

The results in Fig. 3-5 show that for positive τ the CRSS is lower than that for $\tau = 0$, and that in this region the dislocation always glides on the $(\bar{1}01)$ plane. In contrast, negative τ makes the glide on the $(\bar{1}01)$ plane more difficult. With decreasing τ the CRSS increases until the glide plane changes from the $(\bar{1}01)$ plane to the $(0\bar{1}1)$ plane. The above results indicate that although the shear stress perpendicular to the slip direction does not drive directly the screw dislocation to move, it influences the CRSS by altering the dislocation core structure.

In the following, it presents an example on how the core changes in terms of τ, again by comparing the differential displacement plots with and without loading. A special stress tensor applied in the coordinate system where the *y*-axis is normal to the plane defined by the angle χ and the *z*-axis is parallel to the dislocation line was used:

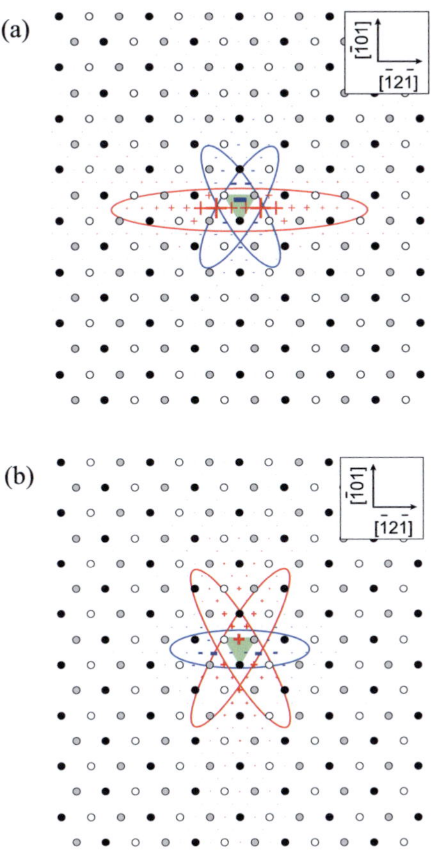

Figure 4-2. Changes of the differential displacements around the core with only pure shear stresses perpendicular to the slip direction applied on the $(\bar{1}01)$ plane for (a) $\tau = +0.02C_{44}$ and (b) $\tau = -0.02C_{44}$. The magnitude of change is multiplied by a factor of 20. The shading is the center of the dislocation core and the ellipses are used to highlight the changes of the distribution of the Burgers vector.

$$\Sigma_{\tau} = \begin{bmatrix} -\tau & 0 & 0 \\ 0 & \tau & 0 \\ 0 & 0 & 0 \end{bmatrix} \tag{4-1}$$

where τ is the magnitude of the shear stress perpendicular to the slip direction, resolved in this orientation as a combination of two normal stresses. For $\chi = 0$, the extension and constriction of the core has to be symmetric with the $(\bar{1}01)$ plane being the mirror plane. The redistribution of the Burgers vector by loading with $\tau = \pm 0.02 C_{44}$ is shown in Fig. 4-2.

For positive τ, the dislocation core extends on the $(\bar{1}01)$ plane and constricts on both $(0\bar{1}1)$ and $(\bar{1}10)$ planes which suggests that the dislocation will move most easily on the $(\bar{1}01)$ plane. On the other hand, for negative τ, the core constricts on the $(\bar{1}01)$ plane and extends on both $(0\bar{1}1)$ and $(\bar{1}10)$ planes. If this spreading is large, the dislocation becomes easier to move on the low stressed $(0\bar{1}1)$ or $(\bar{1}10)$ planes than on the primary $(\bar{1}01)$ glide plane. Hence, one can expect that the subsequent loading by the shear stress parallel to the slip direction will move the dislocation on the $(\bar{1}01)$ plane for $\tau > 0$, while for $\tau < 0$ the preferred glide planes will be $(0\bar{1}1)$ or $(\bar{1}10)$.

Besides the preference for slip on a particular $\{110\}$ plane, the changes in the structure of the dislocation core also suggest how large CRSS is needed to drive the dislocation. For example, let us assume again that the crystal is loaded by the stress tensor Σ_{τ} (Eq. 4-1) defined in the MRSSP coordinate system where the y-axis coincides with the normal to the $(\bar{1}01)$ plane ($\chi = 0°$). If positive τ is applied, the core extension on the $(\bar{1}01)$

plane makes the glide on this plane easier compared to the symmetric core for $\tau = 0$. Hence, one may expect that if $\tau > 0$, CRSS will decrease with increasing τ. This is fully consistent with atomistic result shown in Fig. 3-5(a), in which the CRSS decreases with increasing τ, when $\chi = 0°$. On the other hand, applying a negative τ makes the glide on the $(\bar{1}01)$ plane increasingly more difficult and, at larger negative τ, the $(\bar{1}01)$ glide may be suppressed completely. Instead, the dislocation glide may proceed exclusively on one of the other two $\{110\}$ planes of the $[111]$ zone. This explains why in Fig. 3-5(a) for $\tau < 0$, the CRSS increases first and then the glide plane changes from the $(\bar{1}01)$ plane to the $(0\bar{1}1)$ plane. However, because the shear stress parallel to the slip direction resolved on the inclined planes is only half of the resolved shear stress in the primary $(\bar{1}01)$ plane, larger CRSS for slip of the dislocation is expected for larger negative τ.

With the above analysis, one can now fully understand the results in Fig. 3-5 of Section 3.1.3. For all tested MRSSP orientations χ, with positive τ the CRSS is lower than that with $\tau = 0$ and the dislocation always glides on the $(\bar{1}01)$ plane. The reason is that the positive shear stress perpendicular to the slip direction extends the dislocation core on the $(\bar{1}01)$ glide plane and constricts it on the other two $\{110\}$ planes in the $[111]$ zone. The change of the core promotes the glide of the dislocation on the $(\bar{1}01)$ plane by lowering the corresponding Peierls stress. In contrast, negative τ makes the glide on the $(\bar{1}01)$ plane more difficult. Here the CRSS increases with decreasing τ until the glide plane changes from the $(\bar{1}01)$ plane to the $(0\bar{1}1)$ plane. The reason is the negative shear stress τ perpendicular to the slip direction constricts the dislocation on the $(\bar{1}01)$ plane and extends it on the inclined $(0\bar{1}1)$ and $(\bar{1}10)$ planes. This makes the glide on the $(\bar{1}01)$ plane harder. When τ reaches a critical value, the preferred glide plane changes from $(\bar{1}01)$ to $(0\bar{1}1)$. The changes of the glide plane are indicated by arrows in the bottom of Fig. 3-5. For $\chi = 0°$, $-9°$ and $-19°$ the slip systems $(\bar{1}01)[111]$ and $(0\bar{1}1)[111]$ are activated at the same time for $\tau \approx -0.01C_{44}$. In these cases, the motion of the screw dislocation is composed of alternating jumps on both of these slip systems so that the apparent macroscopic slip takes place on the $(\bar{1}\bar{1}2)$ plane. Such $\{112\}$ slip occurs as well for loading in compression in the $[012]$ direction. Additionally, there exists an extensive experimental evidence of non-crystallographic slip or slip on high index planes for certain uniaxial loadings in pure single crystals of iron and other refractory metals. This macroscopic phenomenon originates from stochastic transitions between the neighbouring $\{110\}$ slip systems. When the CRSS's on the two glide planes are close or identical, both slip systems will be activated at the same time. Thus, the macroscopically observed $(\bar{1}\bar{1}2)$ or non-crystallographic slips result from dislocation moving in a zigzag fashion by elementary steps on both $(\bar{1}01)$ and $(0\bar{1}1)$ planes. However, for large negative τ the extension of the core on the $(0\bar{1}1)$ plane becomes so overwhelming that the dislocation starts to glide only on this plane, despite the Schmid factor being only half of that for the most highly stressed $(\bar{1}01)[111]$ slip system.

For structures with the cubic lattice symmetry, the change of the dislocation core with (χ, τ) should be equal to the change of the core with $(-\chi, \tau)$

by a rotation of π in terms of the $\chi = 0$ diad. It means that if the glide plane is $(0\bar{1}1)$ for (χ, τ), the glide plane for $(-\chi, \tau)$ should be $(\bar{1}10)$. However, this applies only if the effect of the shear stress perpendicular to the slip direction is considered. One can see (cf. Fig. 3-5) that for $\tau < -0.01 C_{44}$ the dislocation always glides on the $(0\bar{1}1)$ plane independent of the sign of χ. The reason is there is a competition between the Schmid factor, the effect of the shear stress perpendicular to the slip direction, and the effect of the twinning-antitwinning asymmetry. For example, when χ is negative, although the core extends more along the $(\bar{1}10)$ plane than the $(0\bar{1}1)$ plane, the Schmid factor is higher on the $(0\bar{1}1)$ plane and the crystal is sheared in the twinning sense. Thus, the Schmid factor and the twinning-antitwinning effect determine together the $(0\bar{1}1)$ as the glide plane. When χ is positive, although the Schmid factor on the $(0\bar{1}1)$ plane is smaller than that on the $(\bar{1}10)$ plane, the shear is in the antitwinning sense around the $(\bar{1}10)$ plane and the core extension on the $(0\bar{1}1)$ plane is larger than that on the $(\bar{1}10)$ plane. In this case, the effects of the perpendicular shear stress and of the twinning-antitwinning asymmetry dominate and the screw dislocation glides on the $(0\bar{1}1)$ plane. Finally, for $\chi = 0$, although the Schmid factor is 2 times higher on the $(\bar{1}01)$ plane than that on the $(0\bar{1}1)$ or $(\bar{1}10)$ planes, the core is constricted on the $(\bar{1}01)$ plane. The extension of the core on the $(0\bar{1}1)$ and the $(\bar{1}10)$ planes is the same, but since the crystal shears in twinning sense, the dislocation prefers to glide on the $(0\bar{1}1)$ plane. In this case, the effect of the shear stress perpendicular to the slip direction and the twinning-antitwinning asymmetry determine the glide plane.

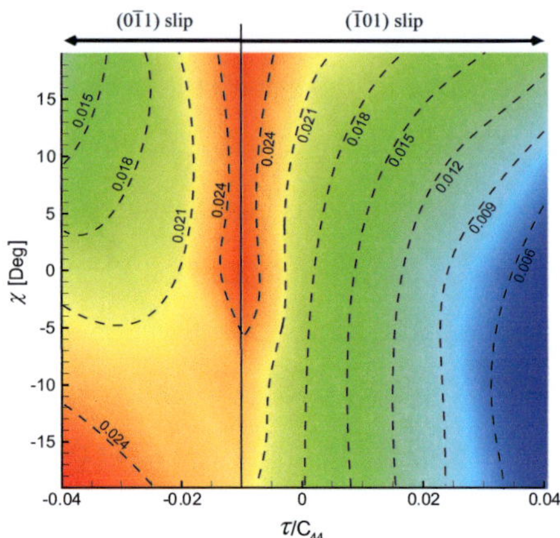

Figure 4-3. The dependence of CRSS on the Schmid factor, the MRSSP angle, χ, and the shear stress, τ, perpendicular to the slip direction.

Fig. 4-3 summarizing the dependence of the CRSS on both χ and τ is the major results of the atomistic calculations. The colour shading in this contour graph represents the change of the CRSS from low (blue) to high (red). As mentioned previously, the CRSS of the screw dislocation is proved atomistically to be determined by the competition of three effects, which are the Schmid factor, the shear stress perpendicular to the slip direction and the twinning-antitwinning asymmetry. One can see in this plot in detail how these effects work competitively. Let us first consider the results for positive τ in the right part of the plot, for which the glide plane is the ($\bar{1}$01) plane. In the horizontal direction, the general decrease of CRSS with increasing τ for any MRSSP orientation originates from the effect of the shear stress perpendicular to the slip direction. In the vertical direction, the variations of CRSS with a constant τ are governed by the twinning-antitwinning asymmetry and the Schmid factor. The CRSS varies strongly only in the antitwinning zone ($\chi > 0$) while in the twinning zone ($\chi < 0$) it remains almost constant. This can be explained qualitatively in the following way. For positive χ, as χ increasing, the Schmid factor increases and the MRSSP rotates towards the anti-twinning plane $(\overline{2}11)$. Both effects increase the CRSS and together cause the strong dependence of CRSS on χ. However, for negative χ, the Schmid factor increases with decreasing χ while the MRSSP rotates towards the twinning plane $(\overline{1}\,\overline{1}2)$. Since they have opposite effect on CRSS and compensate each other, the CRSS remains almost constant as χ varies. As can be seen in the left part of Fig. 4-3, the analysis for negative τ is similar but more complex. Although the three effects together determine $(0\overline{1}1)$ as the glide plane, for the CRSS the competition occurs mainly between the effects of the Schmid factor and the shear stress perpendicular to the slip direction. On one hand, as χ decreases from $+19°$ to $-19°$, the Schmid factor increases on the $(0\overline{1}1)$ plane, indicating that CRSS should decrease for a constant τ. On the other hand, as χ decreases the core extension along the $(0\overline{1}1)$ glide plane is gradually reduced. This constriction has an opposite effect than the Schmid factor leading to an increase of CRSS with decreasing χ. The effect of the Schmid factor dominates only for small $|\tau|$, e.g. $\tau = -0.01C_{44}$, so the CRSS decreases when χ changes from $+19°$ to $-19°$. For larger negative values of τ ($\tau < -0.02C_{44}$), the effect of the shear stress perpendicular to the slip direction prevails and CRSS increases with decreasing χ.

The knowledge of the dependence of CRSS on χ and τ obtained from our atomistic studies enabled us to formulate a phenomenological yield criterion (Eq. 3-3 in Chapter 3) for the non-associated flow in bcc iron. Predictions based on this analytical yield criterion and their comparisons to experimental results are the topics of the following section.

4.2   Yielding of the single crystal by yield criterion

The atomistic simulations provide reliable information about the behaviour of a single $a_0/2[111]$ screw dislocation under stress, but they are too time consuming to investigate large number of loading orientations. Therefore an analytical yield criterion is developed that can determine both quickly and reliably the commencement of the motion of any $a_0/2\langle 111\rangle$ screw dislocation on the 24 slip systems under arbitrary external loadings at 0K.

In order to capture the dependences of the CRSS on both the loading orientation and the non-Schmid stress components, a linear combination of two shear stresses parallel and perpendicular to the slip direction, both resolved in two different $\{110\}$ planes of the $[111]$ zone, was used in the formulation of the yield criterion (Eq. 3-3).

The yield criterion was used to obtain the CRSS vs. τ dependencies for real single crystals of iron. In addition, the yield polygon, which is the yielding surface projected on the CRSS-τ graph for a given MRSSP, was determined. It was found that if the magnitude of the shear stress perpendicular to the slip direction is small, approximately $-0.01C_{44} \leq \tau \leq 0.02C_{44}$, the primary slip system coincides with the most highly stressed $(\bar{1}01)[111]$ system. However, as $|\tau|$ becomes larger, other $\{110\}\langle 111\rangle$ system becomes dominant. If the loading line falls in the close vicinity of the crossing point of two critical lines, both corresponding slip systems will be activated leading to macroscopic slip on the average {211} or some other high-index plane. Since the values of τ at which the plastic deformation of real crystals takes place are bounded by the yield polygon, $|\tau|$ can never be larger than about $0.03C_{44}$.

The yield criterion can be also expressed in a more convenient and efficient form using a tensorial representation [109]. This tensorial form is written as follows:

$$\mathbf{m}^\alpha \mathbf{\Sigma} \mathbf{n}^\alpha + a_1 \mathbf{m}^\alpha \mathbf{\Sigma} \mathbf{n}_1^\alpha + a_2 (\mathbf{n}^\alpha \times \mathbf{m}^\alpha) \mathbf{\Sigma} \mathbf{n}^\alpha + a_3 (\mathbf{n}_1^\alpha \times \mathbf{m}^\alpha) \mathbf{\Sigma} \mathbf{n}_1^\alpha = \tau_{cr}^* \qquad (4\text{-}2)$$

where $\mathbf{\Sigma}$ is the external stress tensor, $\mathbf{m}^\alpha$ is the unit vector of the slip direction, $\mathbf{n}^\alpha$ is the unit vector perpendicular to the reference plane, and $\mathbf{n}_1^\alpha$ the unit vector perpendicular to the {110} plane in the zone of $\mathbf{m}^\alpha$ that makes the angle $-60°$ with the reference plane.

Corresponding to the yield criterion in Eq. 3-3, the first term of the tensorial expression represents the Schmid factor and, together with the second term, they reproduce the twinning-antitwinning asymmetry. The last two terms are projections of $\mathbf{\Sigma}$ on the two inclined {110} planes representing the effect of the shear stress perpendicular to the slip direction. The parameters used in Eq. 4-2 are the same as those in Eq. 3-3. The complete list of the vectors, $\mathbf{m}^\alpha$, $\mathbf{n}^\alpha$ and $\mathbf{n}_1^\alpha$, for all 24 {110}<111> systems has been given in Table. 3-1.

For any applied loading $\mathbf{\Sigma}$ one can assess the activity of each of these 24 reference systems by evaluating the left side of Eq. 4-2. The plastic deformation at 0K then starts when the resolved stress on one of the 24 slip systems reaches τ_{cr}^* as the applied stress tensor $\mathbf{\Sigma}$ increases from zero to the critical value Σ_c^α. The tensorial form of the yield criterion is convenient because it only requires the applied stress tensor Σ defined in the Cartesian coordinate system and no tensorial transformations are required as the evaluation of Eq. 3-3.

4.2.1   Slip behavior under uniaxial loadings

With the help of the yield polygon one can obtain a complete description of the macroscopic yielding behaviour of Fe single crystal. The procedure described in Chapter 3.2.3 can be repeated for any loading using the corresponding stress tensor with only shear stresses parallel and perpendicular to the slip direction in the MRSSP coordinate system. For each loading path such calculations yield four reference systems, each associated with a distinct slip direction, and for every system α a yielding point (a pair of critical shear stresses σ and τ in the MRSSP-τ graph) at which the plastic deformation commences on this system can be determined.

One can now employ the tensorial yield criterion to determine the primary slip systems for loadings in tension and compression along all directions in the standard stereographic triangle for which $(\bar{1}01)[111]$ is the most highly stressed $\{110\}<111>$ slip system.

In Chapter 3.1.1 eight uniaxial loading directions (see Fig. 3-3) were studied by atomistic simulations. For all loadings in tension the most easily activated glide plane is $(\bar{1}01)$ while for compressions it is the $(1\bar{1}0)$ plane. One should note that in the atomistic studies, only the $a_0/2[111]$ screw dislocation exists, indicating that only 6 reference systems, i.e. systems of 1-3 and 13-15, could be operative. However, when considering a real single crystal, all 24 slip systems in Table. 3-1 can be activated since the crystal contains dislocations with all possible Burgers vectors.

Now considering an unit uniaxial applied stress tensor Σ in system α, one can first look up the corresponding vectors $\mathbf{m}^\alpha$, $\mathbf{n}^\alpha$ and $\mathbf{n}_1^\alpha$ defined in Table. 3-1. Then, the left side of Eq. 4-2 can be evaluated and marked as $\tau_{t/c}^{*\alpha}$. According to the yield criterion, the uniaxial tensile/compressive stress for which the system α becomes activated is $\sigma_{t/c}^\alpha = \tau^* / \tau_{t/c}^{*\alpha}$. By repeating this procedure for each of the 24 reference systems one obtain a set of all critical stresses. The actual yield stress inducing the plastic flow is then the smallest of these stresses, i.e. $\sigma_{t/c} = \min(\sigma_{t/c}^\alpha)$, and the corresponding slip system α is the primary slip system.

The primary slip systems predicted by both the Schmid law and the yield criterion are plotted in Fig. 4-4. In the stereographic triangle, regions with different colours indicate different activated slip systems which are labelled with numbers corresponding to Table 3-1.

We consider that a second slip system (labelled with II) can also be activated provided that the required loading in the second system is less than 2% larger than that of the first activated slip system (labelled with I). Such "multi-slip" has been frequently observed in low-temperature deformation experiments on bcc metals [44]. In real situations, the number of the activated slip systems is likely not to be limited by two and the threshold of 2% is only an estimated value for assessing the possibility of the multi-slip phenomenon. Apart from the intrinsic origins of the multi-slip behaviour, this anomalous phenomenon also depends on external loading conditions, e.g., temperature and strain rate.

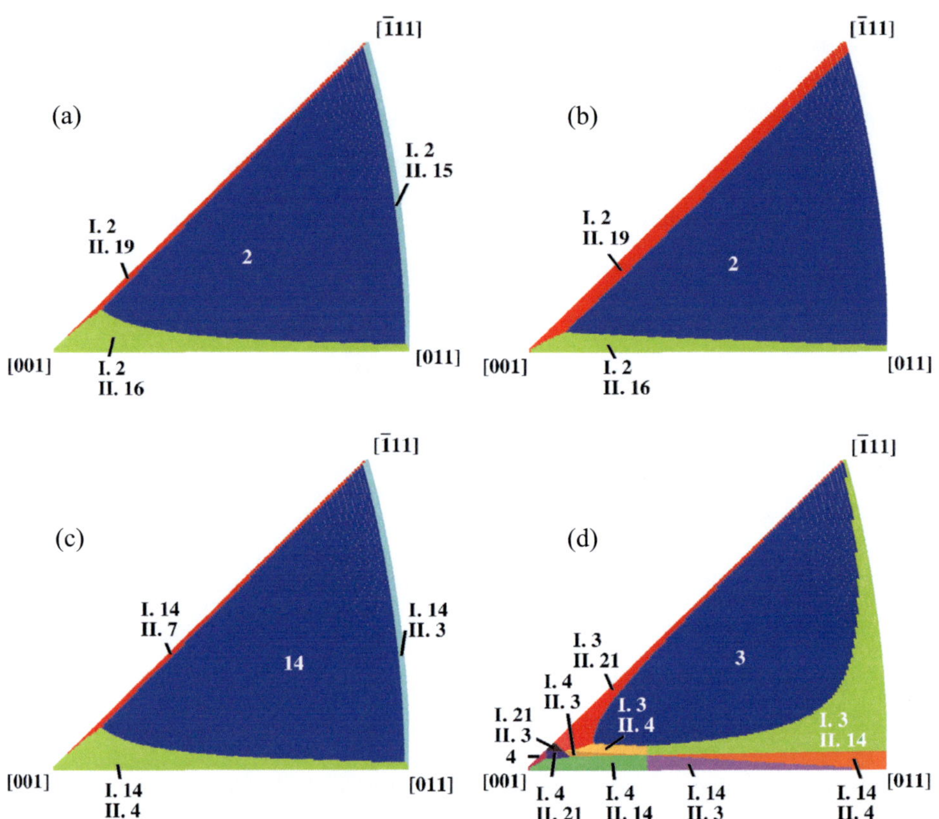

Figure 4-4. Primary slip systems for loadings with all possible orientations within the stereographic triangle in tension (a, b) and compression (c, d) predicted by the Schmid law (a, c) and the yield criterion shown in Eq. 4-2 (b, d). Two slip systems, marked as I and II, are considered to be activated simultaneously when their difference is within 2%.

As illustrated in Fig. 4-4(a) and (b), the predicted primary slip system for tension along any orientation within the standard stereographic triangle, according to both the yield criterion and the Schmid law, is the $(\bar{1}01)[111]$ system with the highest Schmid factor. This slip system has been also found in all atomistic calculations presented in Chapter 3.1.2, where the loadings in tension were applied on the $a_0/2[111]$ screw dislocation for eight different orientations.

The yield criterion and the Schmid law also agree closely in predictions of the second slip system, with only small differences. For example, the Schmid law predicts a multi-slip along the $[011]-[\bar{1}11]$ boundary, while according to the yield criterion only the $(\bar{1}01)[111]$ system is activated.

In the area close to the $[001]-[011]$ boundary, both of the $(\bar{1}01)[111]$ ($\alpha = 2$) and $(\bar{1}0\bar{1})[1\,\bar{1}\,\bar{1}]$ ($\alpha = 16$) slip systems can be activated. This indicates that dislocations with the $a_0/2[111]$ and $a_0/2[1\,\bar{1}\,\bar{1}]$ Burgers vectors will be activated simultaneously. This prediction agrees with experimental findings of Aono and co-workers [68], in which high purity iron single crystal specimens were deformed in tension at very low temperatures, namely 4.2 and 77 K, for various loading orientations. It was found that for most orientations in the stereographic triangle the observed slip system was indeed $(\bar{1}01)[111]$. However, for orientations near to the $[001]-[011]$ side, the multi-slips on both $(\bar{1}01)[111]$ and $(101)[\bar{1}11]$ slip systems were observed.

Another experimental support for our theoretical predictions comes from Spitzig and Keh [188], who deformed high purity Fe single crystals in tension for orientations $0° \leq \chi \leq 20°$ and $\lambda \approx 45°$ between 143-295 K. The observed primary slip occurred on the $(\bar{1}01)$ plane in the $[111]$ zone, but also the second $(101)[\bar{1}11]$ slip system was observed. Since the studied loading orientations fall in the middle of the stereographic triangle, the observation of the second slip system is not on the first sight consistent with our theoretical predictions. The second slip system is predicted to be

Table 4-1. The effective Schmid factor, τ^*, predicted by Schmid law and the yield criterion for loadings in tension and compression along the $[\bar{1}49]$ direction.

Tension				Compression			
Schmid law		Yield criterion		Schmid law		Yield criterion	
α	τ^*	α	τ^*	α	τ^*	α	τ^*
2	1.00	2	1.00	14	1.00	3	1.00
16	0.93	16	0.93	4	0.93	14	0.96
19	0.65	19	0.87	7	0.65	4	0.93
5	0.58	11	0.70	17	0.58	18	0.85
15	0.50	5	0.69	3	0.50	21	0.75

inactive, since the difference of the critical loading between the second and the primary slip systems is larger than 2%. However, $(101)[\bar{1}11]$ is according to our yield criterion indeed the second most favourable slip system. Table 4-1 lists the highest three effective Schmid factors, $\tau_{t/c}^{*\alpha}$, calculated for each system α for uniaxial loadings along the $[\bar{1}49]$ direction with $0° \leq \chi \leq 20°$ and $\lambda \approx 45°$. The effective Schmid factors are normalized by the values corresponding to the system with the highest $\tau_{t/c}^{*\alpha}$. One can clearly see from the table that for tension the critical loading in the second most favourable slip system is only ~7% larger than that in the first slip system. This is larger than our artificially presumed threshold value, 2%, but not impossible to reach in real situations. The reason why it was observed in experiments is likely related to rather high temperature, at which the probability of the activation of the second slip system increases.

The map of the most operative slip systems for compressive loadings predicted by the Schmid law [Fig. 4-4(c)] is identical to that for tensile loadings, only the predicted slip systems for the two loading orientations are conjugate to each other (see Table 3-1). This is because the only difference between tension and compression is that the sense of the shear stress parallel to the slip direction is reversed, so that $(\bar{1}01)[111]$ ($\alpha = 2$) for tension corresponds to $(\bar{1}01)[\bar{1}\bar{1}\bar{1}]$ ($\alpha = 14$) for compression. In contrast to tension, the predictions of the yield criterion for compression [Fig. 4-4(d)] vary considerably with the orientation of the loading axis and are overall much more complex than those of the Schmid law. The most striking difference is that our yield criterion predicts completely different primary slip system ($\alpha = 3$) than the Schmid law ($\alpha = 14$). As found in the atomistic simulations, this result is related to the strong effect of the shear stress perpendicular to the slip direction, which causes the preferential activation of the $\alpha = 3$ over the $\alpha = 14$ slip system, although the latter possesses a much higher Schmid factor. Only close to the [011] corner, $\alpha = 14$ is predicted to be the primary slip system. The dislocation motion on slip systems other than $(\bar{1}01)[111]$ is the well-known anomalous slip observed in most bcc metals [44]. Unfortunately, to our knowledge there are currently no experimental results from low temperature compression testings available to verify our theoretical predictions. However, our study shows the ability of the yield criterion to predict a very complex mechanical behaviour for the iron single crystal under compression.

4.2.2   Yield stress asymmetry in tension and compression

The tension-compression asymmetry in bcc metals was observed experimentally [32, 34, 189-197] in both separate tension/compression tests and successive tension-compression cycles for different loading orientations and temperatures. In most orientations of the standard stereographic triangle, the CRSS for compression was found to be higher than that for tension. This agrees with our atomistic results for iron presented in Chapter 3.1.1. It should be noted that since the tension-compression asymmetry is related to intrinsic properties of screw dislocation, it is obvious at low temperatures but usually becomes negligible as the temperature increases (see later).

Koss concluded already in the 1980's that the tension-compression asymmetry is closely related to the twinning-antitwinning asymmetry [198]. However, based on our atomistic results and the analytical yield criterion, it will show in the following that the tension-compression asymmetry is a consequence of not only the twinning-antitwinning asymmetry but also the strong effect of the shear stress perpendicular to the slip direction, via the so-called strength differential (*SD*) factor:

$$SD = \frac{\sigma_t - \sigma_c}{(\sigma_t - \sigma_c)/2} \tag{4-3}$$

where σ_t and σ_c are the uniaxial yield stresses in tension and compression, respectively. For any orientation of the loading axis these yield stresses can be determined from the yield criterion as described above. Performing this calculation for all orientations of tension/compression axes, a map of the strength differential can be obtained for the whole standard stereographic triangle. This is displayed in Fig. 4-5 by shading the interior of the

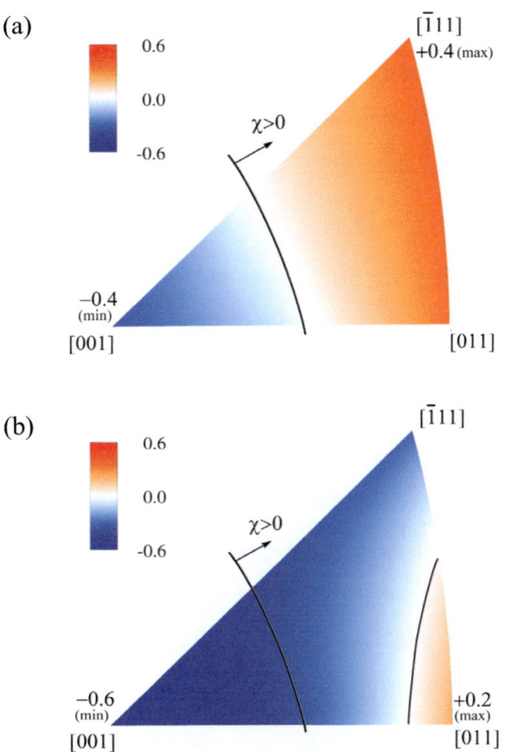

Figure 4-5. Tension-compression asymmetry factor calculated with (a) only twinning-antitwinning asymmetry and (b) the full yield criterion.

standard triangle by the value of *SD* according to Eq. 4-3 (One should note the comparison between loadings in tension and compression in the same direction is regardless of slip systems).

The results drawn in Fig. 4-5(a) were obtained when only the twinning-antitwinning asymmetry was considered by using the first two terms in Eq. 3-3. The distribution of *SD* is in this case anti-symmetric with respect to χ, and SD equals to zero for $\chi = 0$. When χ is positive, the loadings in tension are in the anti-twinning sense while loadings in compression are in the twinning sense. The value of SD is positive in this region and its maximum reaches about 0.4 at $[\bar{1}11]$ corner. Vice versa, SD is negative for $\chi < 0$ where the senses of the twining/anti-twinning are reversed. The minimum value of SD is -0.4 corresponding to uniaxial loading along the [001] direction.

Fig. 4-5(b) contains the SD map calculated with both the effects of the twining-antitwinning asymmetry and the shear stress perpendicular to the slip direction. It is clearly very different from the predictions made with the twinning-antitwinning effect only. Comparing to Fig. 4-5(a), one of the most prominent changes in Fig. 4-5(b) is that the region with positive SD value is greatly reduced to the [011] corner. The complete yield criterion predicts that for most orientations of the uniaxial loadings in the standard stereographic triangle the yield stress for compression is larger than that for tension and thus $SD < 0$. For loading axis close to the $[011]-[\bar{1}11]$ side, the critical stress for tension gradually increases relative to the critical stress for compression and then SD becomes positive. The distribution of SD is no longer anti-symmetric with respect to $\chi = 0$, but the boundary is markedly shifted towards the [011] corner. The maximum positive tension-compression asymmetry lies at the [011] corner with $SD \sim 0.2$, while the minimum value of the strength differential corresponds to the loading axis along [001] where $SD \sim -0.6$.

In experiments, the tension-compression asymmetry in iron single crystal was measured by Zwiesele and Diehl [17]. The sample was uniaxial deformed along the direction for which $11° \leq \chi \leq 12°$ and $\eta \sim 0.72$ (note η is an estimated value since the exact value of λ is not provided in Ref. [17]). The critical resolved shear stresses at the lowest measured temperature of

77 K were ~240MPa for tension and ~280MPa for compression so that SD is about -0.15. As seen in Fig. 4-5, the predicted *SD* values at 0 K for the same loading orientation are +0.17 when only twinning-antitwinning asymmetry is considered and -0.20 using the full yield criterion. The later value agrees very well with the experimental result, and shows the ability of the atomistically-based yield criterion to predict accurately the yield behavior of Fe single crystal at low temperatures.

4.3   **Thermally activated dislocation mobility**

In Chapter 3.3 it developed a link between the behaviour of the $a_0/2<111>$ screw dislocations in bcc iron at 0 K studied by static atomistic simulations and the thermally activated dislocation motion at finite temperatures. The commencement of the dislocation motion is regarded as nucleation and subsequent propagation of kink-pairs that overcome the Peierls barrier with the aid of thermal fluctuations and applied stress. In the line tension model, the Peierls barrier is considered to be dependent on the applied stress tensor and is a function of the MRSSP orientation of the loading and both shear stress components parallel and perpendicular to the slip direction. This dependence has the same origin as that found for the Peierls stress at 0 K by atomistic studies. The crucial connection between the 0 K atomistic data and the thermally activated dislocation motion model is achieved via the construction of the Peierls potential, whose derivative in terms of the dislocation position (Eq. 3-17) gives the Peierls stress.

The Peierls potential is constructed based on the *m*-function (Eq. 3-16), which has the same symmetry as the {111} plane of the bcc lattice. The height of the Peierls potential under zero stress, the twinning-antitwinning asymmetry, and the dependence of the Peierls potential on the shear stress perpendicular to the slip direction are described by parameter functions multiplying the *m*-function. These functions are determined in a self-consistent manner from the analytical yield criterion. The main advantage of the constructed Peierls potential is that it reflects the dependence of the Peierls stress on χ, σ and τ described in Chapter 4.1 and 4.2.

When comparing the stresses obtained by our calculations to those measured in experiments, one should note, deviations of the stresses between the experimental data and the atomistic simulations exist for all bcc metals regardless of the description scheme of the atomic interaction [34, 68, 92, 95, 98, 132, 184-187]. For example, the CRSS of the loading in tension along the [$\bar{1}$49] direction for iron predicted at 0 K in Fig. 3-15 (or obtained by atomistic studies in Chapter 3.1.2) is about 1800 MPa. In contrast, the critical resolved shear stress for approximately the same loading orientation obtained by extrapolating low-temperature experimental measurements of the yield stresses to 0K is between 340 MPa (loaded in tension with axis close to the [$\bar{1}$49] direction but small negative χ, $-8° \leq \chi \leq -6°$) and 390 MPa (loaded in tension with axis close to the [$\bar{1}$49] direction but small positive χ, $0° \leq \chi \leq 8°$) [70]. Hence, the experimentally estimated critical resolved shear stress is ~1/5 of the Peierls stress obtained by the atomistic studies with the similar loading orientation using BOP at 0 K.

Possible explanations that have been discussed in literature are following: (1) a quantum mechanical tunnelling at low temperatures that aids the dislocation to overcome the Peierls barrier [199-202]; (2) a quantum effect on the vibration mode of the dislocation due to discrete energy levels and the zero-point vibration [203-206]; (3) dynamical effects during dislocation motion caused by finite velocity of dislocations, i.e. dislocation inertia [207, 208]; (4) collective effects where a group of dislocations consisting of both non-screws and screws can move at lower applied stress due to mutual interactions [187]. All these effects may contribute to lowering the CRSS at 0 K, thus when comparing the stresses obtained by our calculations to those measured in experiments, it is necessary to reduce the theoretical results by a rescaling factor of ~1/5, with which the predicted σ^* at 0 K, where the required activation enthalpy vanishes, equals to the 0 K yield stress estimated from experiments.

4.3.1   Temperature dependence of the yield stress

The knowledge of the Peierls potential enables us to describe the thermally activated dislocation motion via formation of kink-pairs using standard dislocation models [52, 120-124].

In Chapter 3.3.2 the stress dependence of the activation enthalpy was exemplified for loadings in tension along the $[\bar{1}49]$ direction (see Fig. 3-15). With the determined activation enthalpy, the temperature dependence of the yield stress σ^* for a given slip system at a fixed plastic strain rate $\dot{\gamma}$ can be expressed as [122-124]:

$$H(\sigma) = k_B T \ln\left(\frac{\dot{\gamma_0}}{\dot{\gamma}}\right) \qquad (4\text{-}4)$$

where $H(\sigma)$ is a function of the applied stress. With fixed plastic strain rate $\dot{\gamma}$, the factor $\dot{\gamma}_0$ can be determined from the temperature T_k at which the thermal component of the yield stress vanishes. At this temperature the activation enthalpy is equal to $2H_k$, and thus the prefactor can be obtained from:

$$\ln(\frac{\dot{\gamma}_0}{\dot{\gamma}}) = 2H_k / k_B T_k \qquad\qquad (4\text{-}5)$$

With fixed strain rate, the prefactor may vary with both temperature and stress. However, in certain temperature interval it can be considered as a constant virtually independent of stress and temperature [69, 72, 119]. This approximation is kept also in all our calculations and a constant prefactor is used for a fixed strain rate as in [138].

Experimentally, the temperature dependence of the yield stress in iron single crystal loaded in tension along the axis close to the $[\bar{1}49]$ direction was measured between 4 and 350 K by Brunner and Diehl [69, 70, 209] (See Fig. 4-6). The loading orientation is characterized by $0° \leq \chi \leq 8°$ and $43° \leq \lambda \leq 45°$ with $\eta \sim 0.51$. The plastic strain rate was $\dot{\gamma} = 8.5 \times 10^{-4}$ s^{-1} and the temperature T_k at which the thermal component of the yield stress vanished was approximately 350 K. The corresponding value of the prefactor $\ln(\dot{\gamma}_0 / \dot{\gamma})$ is then ~30 (see [72, 118, 119]). Fig. 4-6 also includes results of Kuramoto et al. [68, 210] for a high purity iron sample deformed in a similar temperature range (4.2 to 300 K) in tension along axis close to the $[\bar{1}49]$ direction with $\chi = 0$ and $\lambda \approx 45°$. The strain rate in this study was $\dot{\gamma} = 1.7 \times 10^{-4}$ s^{-1} which is also similar to the strain rate used in [69, 70, 209]. The corresponding prefactor $\ln(\dot{\gamma}_0 / \dot{\gamma})$ of ~27 agrees well with the

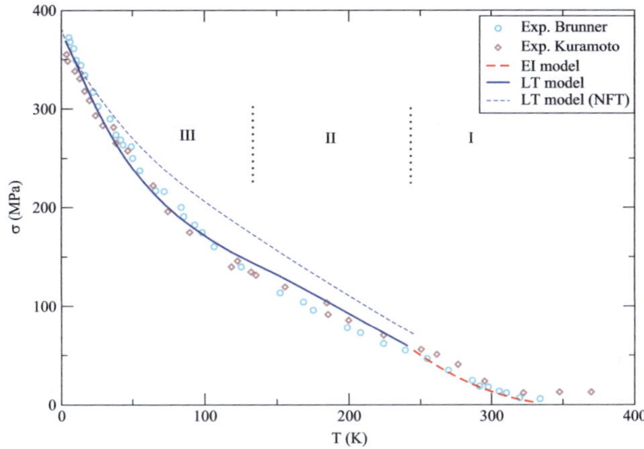

Figure 4-6. Temperature dependence of the yield stress in iron single crystal for loading in tension along the [$\bar{1}$49] direction predicted at high temperatures by the elastic interaction model (red curve), at low temperatures by the line tension model using Peierls potential with flat top (solid blue curve) and without flat top (dashed blue curve), and their comparisons to the experimental data [68-70, 209, 210] (symbols). One should note the calculated shear stresses are rescaled by a factor of ~1/5.

value obtained in studies of Brunner and Diehl. Based on these experimental data, for our calculations it takes $\ln(\dot{\gamma}_0 / \dot{\gamma}) = 30$.

Fig. 4-6 is the comparison between the experimental results and our calculations. As mentioned previously the calculated stresses are rescaled by a factor of ~1/5. The dashed blue curve corresponds to our prediction based on the line tension model at low temperatures. It can be seen in this region obvious differences existing between the prediction and the experimental data. However one should note that the Peierls barrier $V(\xi)$ (Eq. 3-25) derived from the *m*-function has a sharp maximum due to its sinusoidal character. It has been shown in [117, 178-182] that a better agreement between calculated temperature dependence of the yield stress and the experimental data is obtained if the Peierls barrier is flat, i.e. the MEP has a flat plateau instead of the sharp maximum. Mathematically, a truncating of the *m*-function producing the flat plateau, the so-called "flat top", can be obtained [138] by using a flatting operator $\hat{f}$. The important feature of the operator $\hat{f}$ is that it is only applied to every saddle point, i.e. only the sharp maximum of the MEP are removed while the positions and heights of the minima and maxima of *m* remain unaffected. In the current work, a simplified $\hat{f}$ operator which imposes a sharp flatting on the top of the MEP is probed in the following way:

$$V(\xi)=V_f \qquad \text{if } V>V_f \tag{4-6}$$

so that if the energy along the MEP is higher than V_f, it will be set to a constant value V_f. This simple truncating scheme makes the Peierls barrier flat with the critical value V_f determined by:

$$V_f=\frac{V(\xi_1)+V(\xi_2)}{2} \tag{4-7}$$

where

$$\begin{aligned}\xi_1 &= \xi(V_{max})-\Delta\xi \\ \xi_2 &= \xi(V_{max})+\Delta\xi\end{aligned} \tag{4-8}$$

and

$$\Delta\xi=[\xi(V_{max})-\xi_c]\cdot u \tag{4-9}$$

in which $\xi(V_{max})$ and ξ_c are the positions of the dislocation on the transition path with maximum energy and maximum force respectively. The variable *u* in Eq. 4-9 is the only parameter determining the position of the flat top. The choice of *u* is not unique. The basic requirements for its se-

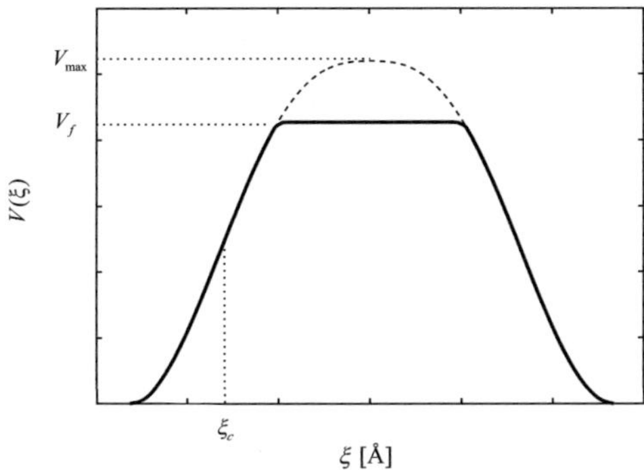

Figure 4-7. The original Peierls barrier $V(\xi)$ without flat top by m-function (dashed curve) and the modified one with flat plateau (solid curve) by the flatting operation in Eq. 4-6.

lection are that, firstly, V_f must be higher than $V(\xi_c)$ and consequently the reasonable value of u is between 0 and 1, and, secondly, the modified MEP should give a better temperature dependence of stress than that without the flat top.

Fig. 4-7 shows the Peierls barrier with the flat top with u equal to 0.65, which was found to be the best choice for Fe. However it is important to emphasize here that the m-function (Eq. 3-16) is just an empirical and therefore possibly crude representation of the shape of the Peierls barrier. The flat-top approximation is based on empirical fitting rather and does not have any solid physical background. A better way to obtain the shape of the Peierls barrier is to determine the MEP directly by means of atomistic calculation using the NEB method to follow the dislocation crossing the barrier. This work is still in progress.

The solid blue curve in Fig. 4-6 based upon the line tension model with flat-topped Peierls potential shows us the temperature dependence of the yield stress of iron at low temperatures. Comparing to the dashed blue curve, it shows that with the flat-top approximation our predictions matches better the experimental data. In this region, the straight dislocation is first shifted away from its minimum position in the Peierls potential by the applied stress and then bows out by thermal fluctuations. When enough energy is provided, the bow-out reaches the critical configuration and continues to expand as a fully developed kink-pair. The activation process depends on the stress components both parallel and perpendicular to the slip direction.

With increasing temperature and decreasing σ^*, the influence of the stress on the shape of the Peierls potential vanishes and the controlling mechanism of the kink-pair formation gradually changes from the line tension model at low temperatures to the elastic interaction model at high temperatures.

For the latter mechanism the fully developed kink pairs are formed by thermal fluctuations. Once enough energy is gained to overcome the energy barrier, the kinks propagate and consequently the screw dislocation moves. This process is determined only by the shear stress component σ^*.

The curves describing the two models intersect at $T \sim 250$ K and stress $\sim$50 MPa, where the activation enthalpies of both models become equal.

In the high temperature region, the yield stress σ^* continues to decrease with increasing temperature until it vanishes at $T \sim 350$ K. Above this temperature only the constant athermal stress $\bar{\sigma}$ is sufficient to move the dislocations.

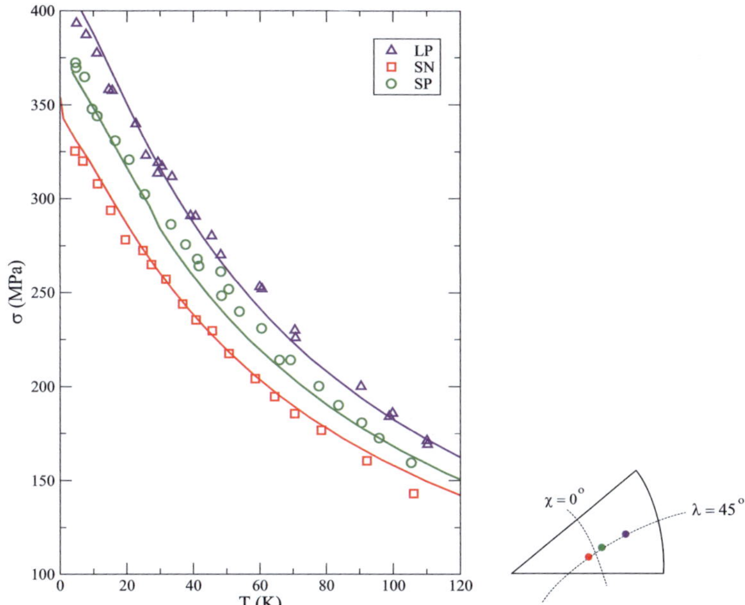

Figure 4-8. Temperature dependences of the twinning-antitwinning asymmetry in iron single crystal predicted by the line tension model (curves) at low temperatures and their comparisons to the experimental data [69, 70] (symbols), by loadings in tension along three orientations with $18° \leq \chi \leq 22°$ (LP), $0° \leq \chi \leq 8°$ (SP), $-8° \leq \chi \leq -6°$ (SN) and the same η value. One should note the calculated shear stresses are rescaled by a factor of $\sim 1/5$.

4.3.2 Temperature dependence of the twinning-antitwinning asymmetry

The general formulation of the Peierls potential developed above provides also a natural description of the twinning-antitwinning asymmetry at finite temperatures. The dependence of the yield stress on the orientation of the

MRSSP χ at different temperatures can be compared directly to available experimental results.

The twinning-antitwinning asymmetry in iron single crystal loaded in tension was experimentally measured between 4 and 110 K by Brunner and Diehl [69, 70] (See Fig. 4-8). Three different angles χ between the primary slip plane $(\overline{1}01)$ and the MRSSP were selected, namely $18° \leq \chi \leq 22°$, $0° \leq \chi \leq 8°$ and $-8° \leq \chi \leq -6°$ for samples marked as large positive (LP), small positive (SP), and small negative (SN), respectively. For all samples, the angle λ between the tension direction and the $[111]$ slip direction was $43° \leq \lambda \leq 45°$, which corresponds to $\eta \sim 0.51$. The plastic strain rate was $\dot{\gamma} = 8.5 \times 10^{-4}$ s^{-1} and the corresponding value of the prefactor was $\ln(\dot{\gamma}_0 / \dot{\gamma}) = 30$ [119].

Calculations with the same loading orientations as those in the above experiments are performed. The curves in Fig. 4-8 are our predictions based on the line tension model at low temperatures (One should note the calculated stresses are rescaled by a factor of $\sim 1/5$). We see that in the whole temperature range the yield stress with positive χ is higher than that with negative χ at the same temperature, while the values for $\chi = 0$ lie inbetween. This agrees with our atomistic simulations shown in Fig. 3-3 in Chapter 3.1.1 where the same twinning-antitwinning tendency was obtained at 0 K.

The result also shows an excellent agreement between our predictions and the experimental data, which demonstrates the ability of our model to predict accurately the twinning-antitwinning asymmetry. In addition, it can be seen from both the experimental data and our theoretical predictions that the twinning-antitwinning asymmetry decreases with increasing temperature. The magnitude of the twinning-antitwinning asymmetry is therefore inversely proportional to T in this temperature region. It can be expected that the twinning-antitwinning asymmetry vanishes when the controlling mechanism of the kink-pair formation changes from the low temperature model to the high temperature model at ~250K where the mechanism of fully developed kink-pair becomes applicable. Above this temperature, the energy barrier does not change significantly during the loading, and the activation enthalpy becomes only a function of the shear stress σ^* projected on the slip plane (Eq. 1-4).

4.3.3 Temperature dependence of the tension-compression asymmetry

Similar to the twinning-antitwinning asymmetry, the tension-compression asymmetry is usually obvious at low temperatures but negligible at room temperature [44]. For example, the tension-compression asymmetry was observed to decrease smoothly with increasing temperature for a loading orientation in the center of the stereographic triangle in single crystal niobium despite a transition in the slip behaviour [197].

The tension-compression asymmetry in iron single crystal was measured at temperatures between 77 and 410 K by Zwiesele and Diehl [17] (See Fig. 4-9). The sample was uniaxial deformed along the direction for which $11° \leq \chi \leq 12°$ and $\eta \sim 0.72$ (η is an estimated value since the exact value of λ was not provided in Ref. [17]). The plastic strain rate was $\dot{\gamma} = 5.6 \times 10^{-4}$ s^{-1} and the corresponding prefactor $\ln(\dot{\gamma}_0 / \dot{\gamma}) = 29 \pm 1$. In this temperature range, the experimentally observed glide system for both tension and compression is the $(\bar{1}01)[111]$ slip system. Recalling Fig. 4-4(d) in Chap-

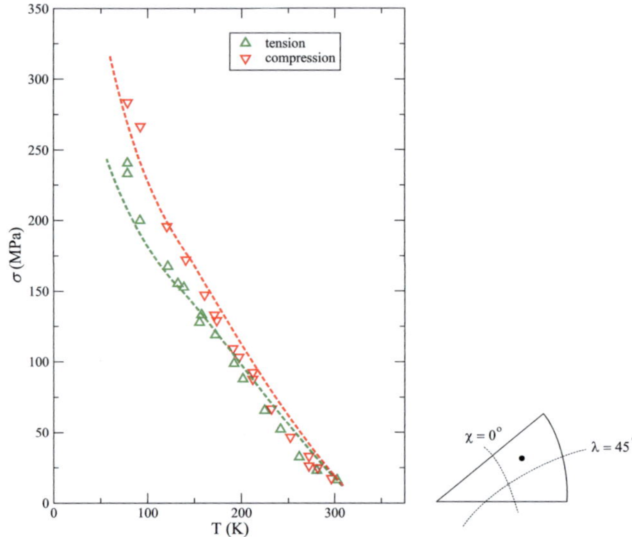

Figure 4-9. Temperature dependence of tension-compression asymmetry in iron single crystal predicted by the line tension model (curves) at low temperatures and its comparison to the experimental data [17] (symbols), by loadings in tension and compression along orientation with $11° \leq \chi \leq 12°$ and $\eta \sim 0.72$. One should note the calculated shear stresses are rescaled by a factor of $\sim 1/5$.

ter 4.2.1, the predicted slip system for compression with the same loading orientation at 0 K is however $(1\bar{1}0)[111]$. The explanation of this discrepancy lies in an increasing probability for the activation of the $(\bar{1}01)[111]$ slip system, which possesses the second lowest critical loading stress at 0 K (Table 4-1), with increasing temperature. Such temperature dependent slip behaviour will be discussed in detail in the next section.

Fig. 4-9 presents our theoretical predictions, with a rescaling factor of $\sim 1/5$ for the calculated stresses, and their comparison with the experimental data. It is seen that the model reproduces closely the experimental results, and predicts a significant tension-compression asymmetry mainly at low temperatures. The magnitude of the difference decreases with increasing temperature or decreasing σ^*, and vanishes at $T > 250K$, at which the saddle-point configuration for the formation of kink-pairs changes from the bow-out to a pair of fully formed kinks. As already discussed in Chapter 4.2.2, the tension-compression asymmetry is the consequence of the twinning-antitwinning asymmetry and the effect of the shear stress perpendicular to the slip direction [101, 109]. With increasing temperature, both effects diminish and, consequently, also the tension-compression asymmetry decreases until it vanishes completely.

The tension-compression asymmetry predicted by our model shows again a very good agreement with the experimental data. It indicates that the constructed two-dimensional Peierls potential is capable of describing the temperature dependence behaviour originating from the shear stresses perpendicular to the slip direction, which is one of the most important factors distinguishing bcc and close-packed metals.

4.3.4   Temperature dependence of the slip system

We have illustrated in Chapter 3.2.3 that one can predict the glide of screw dislocations at 0 K on all 24 slip systems in bcc Fe single crystal. The dislocation may glide on the slip plane with lower Schmid factor, owing to the shear stress perpendicular to the slip direction. However, as shown in the previous section, this effect decreases with increasing temperature. This indicates that the anomalous slip at low temperatures can be replaced by the normal slip on the glide plane with higher Schmid factor at high temperatures. Thus, when the temperature is taken into account, the yield

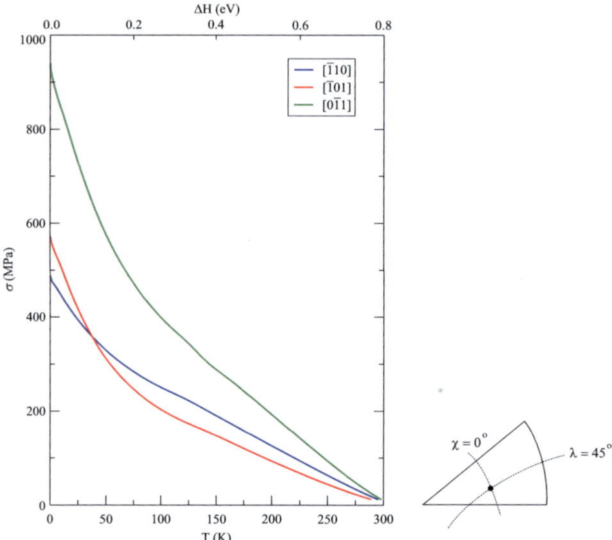

Figure 4-10. Temperature dependence of the activated slip system for loadings in compression along the [$\bar{1}$49] direction in iron single crystal. The temperature dependences of the critical yield stress for the $a_0/2$[111] screw dislocation were predicted by the line tension model on three {110} slip planes of the [111] zone. One should note the calculated shear stresses are rescaled by a factor of ~1/5.

criterion of Eq. 3-3 in terms of stress only is no longer adequate to determine the slip behaviour of the $a_0/2$<111> dislocations in bcc iron. In the current section it will show in detail how the slip system varies with temperature.

At finite temperatures, the yielding occurs firstly on the slip system with the lowest activation enthalpy. Since the Peierls potential $V_\alpha(x,y)$ depends strongly on the loading orientation, its variations are not the same for different slip systems characterized by a unique set of values (χ_α, σ_α, τ_α). Consequently, also the dependencies of the activation enthalpies on the applied loading can be different, and therefore different slip systems with the lowest activation enthalpy may be activated as temperature changes. An illustrative example of such behaviour is the loading in compression along the $[\bar{1}49]$ direction. Table 4-1 in Section 4.2.1 listed the three largest effective Schmid factors $\tau_{t/c}^{*\alpha}$ among all possible slip systems. One can see that at 0 K the critical loading for the second most favourable slip system, $(\bar{1}01)[\bar{1}\,\bar{1}\,1]$ ($\alpha = 14$), is only ~4% larger than that for the primary slip system, $(\bar{1}10)[\bar{1}\,\bar{1}\,\bar{1}]$ ($\alpha = 3$).

Fig. 4-10 shows the temperature dependences of the CRSS for compression along the $[\bar{1}49]$ direction ($\chi = 0$ and $\lambda = -0.5$) for the screw dislocation with $a_0/2[111]$ Burgers vector on the three glide planes in the $[111]$ zone. To consistent with previous discussion, the calculated stresses are again rescaled by a factor of ~1/5.

At very low temperatures close to 0 K, the primary slip system is indeed $(\bar{1}10)[\bar{1}\,\bar{1}\,\bar{1}]$ ($\alpha = 3$) as predicted by our yield criterion [cf. Fig. 4-4(d)] . However, as the temperature increases to ~40K, the slip plane with the lowest critical stress changes from $(\bar{1}10)[\bar{1}\,\bar{1}\,\bar{1}]$ ($\alpha = 3$) to $(\bar{1}01)[\bar{1}\,\bar{1}\,1]$ ($\alpha = 14$). This change of the primary slip system helps to resolve the apparent disagreement between our theoretical predictions based on the yield criterion and the experimental observations. As already mentioned, Zwiesele and Diehl [17] found that the slip above 77 K indeed occurs on $(\bar{1}01)[\bar{1}\,\bar{1}\,1]$ ($\alpha = 14$) and not on $(\bar{1}10)[\bar{1}\,\bar{1}\,\bar{1}]$ ($\alpha = 3$) for both tension and compression, in perfect agreement with results shown in Fig. 4-10. As discussed in Chapter 4.2.1, similar discrepancies, where the experimentally observed slip system is the one with the second lowest critical yield stress, exist for other tensile loadings. The explanation of these apparent discrep-

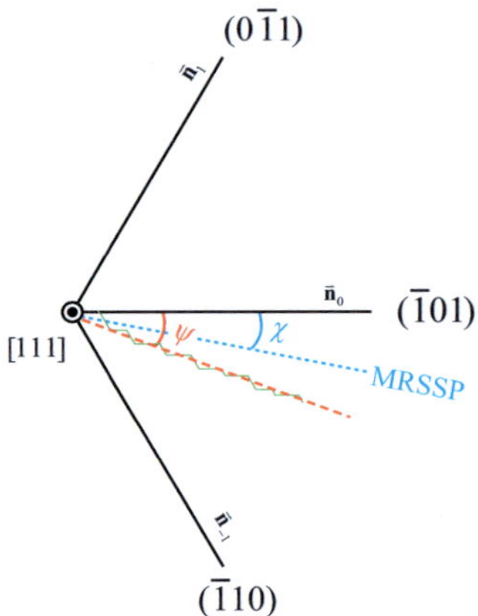

Figure 4-11. Zig-zag slip (green lines) of the $a_0/2[111]$ screw dislocation glide macroscopically on a high-index slip plane (red) in the [111] zone under compression.

ancies is likely equivalent to that presented here. The predictions in Fig. 4-4 are based on the yield criterion developed for 0 K, while experimental observations are always performed at finite temperatures. As the temperature increases, the effect of the shear stress perpendicular to the slip direction decreases and the slip changes from the anomalous slip plane to the normal slip plane, which has usually a higher Schmid factor than the former. It is therefore necessary to consider the effect of temperature when comparing the theoretical predictions with experiments as changes of the slip systems with temperature have been observed frequently in bcc metals [44, 68, 188]. If the activation enthalpies for two slip systems are similar (for instance, in the vicinity of the crossing of the curves in Fig. 4-10), both of these slip systems can be activated at the same time. In this case, the macroscopically observed slip plane could be a {211} plane or any high index plane in the zone of the [111] slip direction.

Finally, the total plastic strain rate can be determined by summation of contributions of dislocations on all possible glide planes (cf. Eq. 3-29). Thus for the $a_0/2[111]$ dislocation, the total velocity can be written as:

$$\mathbf{v} = \mathbf{v}_1 \bar{\mathbf{n}}_1 + \mathbf{v}_0 \bar{\mathbf{n}}_0 + \mathbf{v}_{-1} \bar{\mathbf{n}}_{-1} \tag{4-10}$$

where $\bar{\mathbf{n}}_i$ are the unit vectors of the three glide planes in the $[111]$ zone (Fig. 4-11) and $\mathbf{v}_i$ are the corresponding velocities:

$$\mathbf{v}_i = v_0 \exp[-\frac{\Delta H_i}{k_B T}] \tag{4-11}$$

Eq. 4-10 enables us to determine the dependence of the angle ψ between the $(\bar{1}01)$ plane and the macroscopic slip plane as a function of temperature. This dependence is plotted in Fig. 4-12.

At temperatures below 25 K, the dislocation velocity in Eq. 4-10 is entirely determined by $\mathbf{v}_{-1}$ along the $(\bar{1}10)$ plane of the lowest activation enthalpy ΔH_{-1} so that $\psi = -60°$. Since the activation enthalpy appears in Eq. 4-11 in exponent, the contribution from the other two slip systems with higher ΔH_i can be safely neglected (see Fig. 4-10).

In the transition region between 25 and 40 K the $(\bar{1}10)$ plane possesses still the lowest activation enthalpy, but the difference between the $(\bar{1}10)$ and $(\bar{1}01)$ planes becomes smaller. Both contributions need to be included and consequently ψ starts increasingly deviating from $-60°$. The critical point where the activation enthalpies for the $(\bar{1}10)$ and $(\bar{1}01)$ planes become equal occurs at T $\sim$ 40K. At this temperature both planes are equally active and the macroscopic slip plane is the $(\bar{2}11)$ plane.

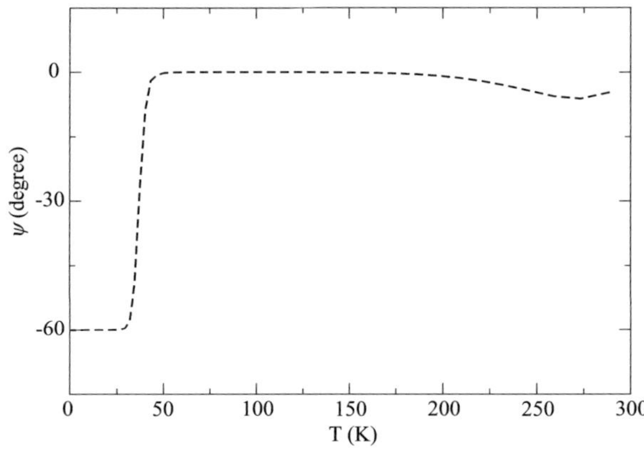

Figure 4-12. Temperature dependence of the angle ψ between the macroscopic slip plane and the $(\bar{1}01)$ plane for the $a_0/2[111]$ screw dislocation under compression.

For temperatures between 40 and 55 K the $(\bar{1}01)$ plane has the lowest activation enthalpy, but the value is still comparable to that of the $(\bar{1}10)$ plane. Thus, ψ gradually rotates from $-30°$ to $0°$. Between 55 and 175 K the $(\bar{1}01)$ plane completely dominates and $\psi = 0°$. The situation above 175 K is more complex. As the temperature increases above 175 K, the difference between the two lowest ΔH_i of the $(\bar{1}10)$ and $(\bar{1}01)$ planes reduces again (see Fig. 4-10), so that the contribution from the $(\bar{1}10)$ plane becomes non-negligible. The macroscopic slip plane therefore deviates somewhat from $0°$ in this temperature region. When the temperature is larger than 250K, the activation enthalpies on all three planes converge to close values. When the contribution from the three glide planes are the same, the average velocity vector results in $\psi = 0°$ and the average slip plane rotates back-towards the $(\bar{1}01)$ plane.

Fig. 4-12 reveals that the macroscopic slip plane can vary considerably depending on the loading orientation and temperature. According to the Schmid law, the slip of the $a_0/2[111]$ screw dislocation should always occur on the plane with the highest resolved shear stress, i.e. the $(\bar{1}01)$ plane. However, our predictions show the possible occurrence of the anomalous slip [44, 77, 78] at low temperatures. As the temperature increases, the effect of the perpendicular shear stress is reduced, and the Schmid factor gradually dominates. Consequently, the macroscopic slip plane rotates from the anomalous $(\bar{1}10)$ slip plane to the normal $(\bar{1}01)$ slip plane. The slip behaviour of real single or polycrystalline materials at finite temperatures can be still much more complex, since dislocations with other Burgers vector can be also involved.

5   Summary and outlooks

The main goal of this thesis was to study the properties of the $a_0/2<111>$ screw dislocations in bcc α-iron, and to establish a link between the microscopic behavior of these defects and the macroscopic plasticity.

It started with investigation of the dislocation core structure by means of static atomistic simulations. The inter-atomic interactions were described by the recently developed magnetic bond-order potential [100] that is able to describe correctly both the angular character of bonding and the magnetic interactions in iron. Despite its quantum mechanical origin, BOP is not limited by the periodic boundary conditions and is sufficiently computationally efficient for the modeling of dislocations. The core structure of the $a_0/2<111>$ screw dislocation with this magnetic BOP was found to be non-degenerate and invariant with respect to both the [111] threefold axis and the [$\bar{1}$01] diad. This is the core structure found in all DFT calculations for bcc metals [93-97], in contrast to the degenerate core structures obtained by most empirical potentials [54, 84-87]. BOP is also able to predict the Peierls barrier for screw dislocation moving between two neighboring stable sites in quantitative agreement with DFT calculations [157] as well as experimental estimations [158] (Fig. 2-1).

In the next step, the focus is on the behavior of the 1/2[111] screw dislocation under externally applied stress. In Chapter 3.1 it studied loadings by pure shear stress parallel to the slip direction. The loadings were applied on different MRSSP defined by the angle χ between the MRSSP and the $(\bar{1}01)$ plane. The dependence of the CRSS on χ in Fig. 3-3 shows the deviation of the CRSS between the atomistically obtained data and the predictions from the Schmid law. This twinning-antitwinning asymmetry was observed in experiments (see for example [44, 68]). For all loadings with pure shear stress parallel to the slip direction the slip is always observed to be on the $(\bar{1}01)$ glide plane which has the highest Schmid factor within $-30° \leq \chi \leq 30°$. It is also the glide plane for the $1/2[111]$ screw dislocation in molybdenum and tungsten under the same loadings [101].

Beside the pure shear stresses, the uniaxial loadings in both tension and compression were performed. Our results show that the CRSS for tension is always lower than that for compression in the same loading direction, and that the CRSS for pure shear with the same MRSSP lies in between of the two. This so-called tension-compression asymmetry observed in experiments is a consequence of the non-planar core structure. The origins of the tension-compression asymmetry were analyzed by further calculations in which a special stress tensor (Eq. 4-1) with only shear stresses perpendicular to the slip direction was applied. The differential displacement plots in Fig. 4-2 show that although it does not drive directly the screw dislocation to move, the shear stress perpendicular to the slip direction changes the symmetry of the core and makes the dislocation either easier or harder to slip on different $\{110\}$ planes in the $<111>$ zone. The outcome of our static atomistic simulations showed that the complex dependence of the CRSS is governed by three factors, namely, the Schmid factor, the shear stress perpendicular to the slip direction, and the twinning- antitwinning asymmetry.

Based on the atomistic results, a description of the macroscopic yielding of single crystals containing $a_0/2<111>$ screw dislocations with all possible Burgers vectors was formulated in terms of an analytical yield criterion. The motion of the screw dislocations is considered to be triggered once the external loading satisfies the yield criterion in any of the 24 slip systems. In order to capture the dependences of the CRSS on both the MRSSP orientation and the non-Schmid stress components, a linear combination of two shear stresses parallel to and two shear stresses perpendicular to the slip direction, both resolved in two different $\{110\}$ planes of the [111] zone, was used to construct the analytical yield criterion (Eq. 3-3), following the studies in [109]. The obtained yield criterion was shown to reproduce closely the atomistic data for not only the CRSS of the glide on the primary $(\bar{1}01)$ plane but also that of the anomalous slip on the other $\{110\}$ planes. In addition, the yield criterion was used to obtain the yield polygon, which is the yielding surface projected on the CRSS-τ graph for a given MRSSP. This yield polygon shows a more complex deformation behavior carried by the $a_0/2<111>$ screw dislocations in single crystal iron than yield polygon derived from the Schmid law.

A convenient tensorial representation of the yield criterion was then utilized to determine the first activated slip systems under a given uniaxial loading for all orientations in the stereographic triangle. The results presented in Fig. 4-4 show that the primary slip system for most tensile loadings is the $(\bar{1}01)[111]$ slip system, which possesses the highest Schmid factor. Possible secondary slip systems exist only in the vicinity of the $[001]-[\bar{1}11]$ and $[001]-[011]$ borders of the stereographic triangle. The theoretical predictions agree with available experiments [68]. While the predictions for tension are similar to those of the Schmid law, the slip systems for compressions determined using the yield criterion are much more complex than those using Schmid law. Most striking is the prediction of different primary slip systems in the central region of the stereographic triangle, which is clearly a consequence of the strong effect of the shear stress perpendicular to the slip direction. Furthermore, the first activated slip system varies considerably with the orientation of the loading axis, showing a large complexity of the deformation behavior of iron single crystals in compression.

Based on the calculations above, the tension-compression asymmetry is analyzed for all orientations of the loading axis in the stereographic triangle using the strength differential. It showed that the tension-compression behavior originates mainly due to the effect of the shear stress perpendicular to the slip direction. Our results again agree well with experiments according to which in most regions of the stereographic triangle the critical loading for compression is higher than that for tension [44].

The third main topic of the thesis was to develop a link between the glide of the $a_0/2\langle 111\rangle$ screw dislocations in bcc iron at 0 K studied by the static atomistic simulations and the thermally activated glide of dislocations at finite temperatures. For the latter the commencement of the dislocation motion is regarded as a nucleation and subsequent propagation of kink-pairs, which overcome the Peierls barrier under the effect of the applied stress and the aid of thermal fluctuations. The Peierls barrier is considered to be dependent on the applied stress tensor and is a function of the MRSSP orientation of the loading and both shear stress components parallel and perpendicular to the slip direction. This dependence has the same origin as that found for the Peierls stress in the atomistic studies for a single $a_0/2\langle 111\rangle\{110\}$ dislocation. The connection between them is that the

Peierls stress along the glide plane should equal to the derivative of the Peierls potential in terms of the dislocation position.

The Peierls potential was constructed based on the m-function [137, 138], which satisfies the symmetry of the $\{111\}$ plane in the bcc lattice. The heights of the Peierls potential under zero stress as well as its changes under general applied stress were described by parameter functions multiplying the m-function. These parameter functions were determined in a self-consistent manner using the yield criterion. In this way, the Peierls potential inherited all the properties of the yield criterion i.e., its dependences on the Schmid factor, the twinning-antitwinning asymmetry, and the shear stress perpendicular to the slip direction.

Using the atomistically consistent Peierls potential, the thermally activated dislocation motion via formation of kink-pairs is treated using standard dislocation models [52, 120-124]. The application of these models leads to a correct description of the temperature dependence of the yield stress. The dependence of the twinning-antitwinning asymmetry on temperature is also predicted successfully, and compares well with the experimental data for three loading orientations in tension with different χ but the same η value (Fig. 4-8). The twinning-antitwinning asymmetry decreases with increasing temperature, and is expected to vanish when the activation model changes from the bow-out mechanism (line tension model) to the fully formed kink-pairs (elastic interaction model). There is also an excellent agreement between our predictions and experimental results [17] for the tension-compression asymmetry. Similar to the twinning-antitwinning asymmetry, the predicted magnitude of the tension-compression asymmetry also decreases with increasing temperature since the activation enthalpies for tension and compression converge. This occurs when the kink-pair formation mechanism changes from the low temperature model to the high temperature model at which fully developed kink pairs dominate.

It has been mentioned that the experimentally observed slip system for loading in compression along the direction studied in [17] is the $(\bar{1}01)[111]$ slip system, which is different to the predicted slip systems for the same loading orientation at 0 K using the yield criterion (Fig. 4-4). The reason is that the predictions in Fig. 4-4 are based on the yield criterion which is developed for 0K while experiments are performed at finite temperatures. When considering the single crystal with dislocations of all possible Burgers vectors, the yielding happens on the slip system with the lowest activation enthalpy. Since the activation enthalpy varies with temperature, the slip system may also change. For the compression along the $[\bar{1}49]$ direction, the predicted slip system at low temperatures close to 0 K is $(\bar{1}10)[\bar{1}\,\bar{1}\,\bar{1}]$. However, as temperature increases to ~40 K, the slip system changes to $(\bar{1}01)[111]$, in good agreement with experimental observations.

As mentioned above, it is necessary to apply a scaling factor of 1/5 when comparing the CRSS obtained from our atomistic calculations to that estimated from low-temperature experimental measurements of the yield and flow stresses by extrapolating to 0 K. This discrepancy between the experimental and atomistic results exists for all bcc metals regardless the description of the atomic interaction [34, 68, 92, 95, 98, 132, 184-187]. A satisfactory explanation is still lacking. However, since this constant factor only rescales the absolute CRSS values, it should not alter any of the qualitative results presented here.

Finally, the author believes this work reached the goal to describe the plastic deformation behavior of iron single crystal at finite temperatures based on the microscopic properties of the screw dislocations. However, since the atomistic simulations of dislocation are limited by short accessible length and time scales, there is still room for other modeling schemes. Probably the most natural mesoscopic approach is the Discrete Dislocation Dynamics (DDD) that is able to follow evolution of dislocation ensembles based upon the single dislocation mobility laws to study the macroscopic deformation behavior in metals in real time. The dislocation mobility laws in DDD models is currently based almost exclusively on the Kocks law [120, 139], which describes the activation enthalpy as a function of the resolved shear stress by fitting the parameters to experimental results. This mobility law does not reflect the non-Schmid effects, e.g., the twinning-antitwinning and tension-compression asymmetries observed in both experiments and atomistic studies. In the current work, it established a bottom-up model which can deliver information about dislocation mobilities directly from the microscopic level. The results obtained in this work therefore can be utilized in the future as input data in higher level modeling schemes such as DDD simulations.

References

1.	Schmid, E. *Zn - normal stress law*. in *1st International Congress for Applied Mechanics*. 1924. Delft: Technische Boekhandel en Drukkerij J.Waltman Jr.

2.	Schmid, E. and Boas, W., *Kristallplastizität unter besonderer berücksichtigung der metalle*. 1935, Berlin: Springer.

3.	Seeger, A., in *Encyclopedia of physics*, Flugge, S., Editor 1958, Springer: Berlin. p. 1.

4.	Taylor, G. I. and Elam, C. F., *The distortion of iron crystals*. Proc. Roy. Soc. Lond. A, 1926. **112**: p. 337-361.

5.	Taylor, G. I., *The deformation of crystals of beta-brass*. Proc. Roy. Soc. Lond. A, 1928. **118**: p. 1-24.

6.	Waseda, Y. and Isshiki, M., *Purification process and characterization of ultra purity metals: Application of basic science to metallurgical processing*. 2002, Berlin: Springer.

7.	Cox, J. J., Horne, G. T., and Mehl, R. F., *Slip, twinning and fracture in single crystals of iron*. Trans. ASM, 1957. **49**: p. 118-131.

8.	Hull, D., *Orientation and temperature dependence of plastic deformation processes in 3.25% silicon iron*. Proc. R. Soc. Lond. A., 1963. **274**: p. 5-20.

9.	Matsui, H., et al., *Mechanical properties of high purity iron*. Trans. JIM, 1978. **19**: p. 163-170.

10.	Nine, H. D., *Asymmetry of slip in fatigue of iron single crystals*. Scr. Metall., 1970. **4**: p. 887-891.

11.	Šesták, B. and Zárubová, N., *Asymmetry of slip in Fe-Si alloy single crystals*. Phys. Stat. Sol. B, 1965. **10**: p. 239-250.

12.	Šesták, B., Zárubová, N., and Sladek, V., *Slip planes in Fe-3 percent Si single crystals deformed at 77 degrees K*. Can. J. Phys., 1967. **45**: p. 1031.

13.	Steijn, R. P. and Brick, R. M., *Flow and fracture of single crystals of high purity ferrite*. Trans. ASM, 1953. **46**: p. 1406-1448.

14.	Tadami Taoka, Shin Takeuchi, and Furubayashi, E., *Slip systems and their critical shear stress in 3% silicon iron*. J. Phys. Soc. Jpn., 1964. **19**: p. 701-711.

15.	Taoka, T., Takeuchi, S., and Furubayashi, E., *Slip systems and their critical shear stress in 3% silicon iron.* J. Phys. Soc. Jpn., 1964. **19**: p. 701-711.

16.	Tseng, D. and Tangri, K., *Temperature and strain rate sensitivities of flow stress for the high purity AISI iron.* Scr. Metall., 1977. **11**: p. 719-723.

17.	Zwiesele, S. and Diehl, J. *Temperature and strain rate dependence of the macro yield stress of high purity iron single crystals.* in *5-th Internat. Conf. Strength of Metals and Alloys.* 1979. Pergamon Press.

18.	Altshuler, T. L. and Christian, J. W., *The mechanical properties of pure iron tested in compression over the temperature range 2 to 293 degrees K.* Phil. Trans. R. Soc. Lond. A., 1967. **261**: p. 253-287.

19.	Takeuchi, S., Furubayashi, E., and Taoka, T., *Orientation dependence of yield stress in 4.4% silicon iron single crystals.* Acta Metall., 1967. **15**: p. 1179-1191.

20.	Tomalin, D. S. and McMahon Jr, C. J., *On the plastic asymmetry in iron crystals.* Acta Metall., 1973. **21**: p. 1189-1193.

21.	Zarubova, N. and Sestak, B., *Plastic deformation of Fe-3% Si single crystals in the range from 113 to 473 K - 1. Thermally activated plastic flow.* Phys. Status Solidi A, 1975. **30**: p. 365-374.

22.	Mitchell, T. E., Foxall, R. A., and Hirsch, P. B., *Work-hardening in niobium single crystals.* Philos. Mag., 1963. **8**: p. 1895-1920.

23.	Mitchell, T. E. and Spitzig, W. A., *Three-stage hardening in tantalum single crystals.* Acta Metall., 1965. **13**: p. 1169-1179.

24.	Bowen, D. K., Christian, J. W., and Taylor, G., *Deformation properties of niobium single crystals.* Can. J. Phys., 1967. **45**: p. 903-938.

25.	Foxall, R. A., Duesbery, M. S., and Hirsch, P. B., *The deformation of niobium single crystals.* Can. J. Phys., 1967. **45**: p. 607-629.

26.	Duesbery, M. S., *The influence of core structure on dislocation mobility.* Philos. Mag., 1969. **19**: p. 501-526.

27.	Duesbery, M. S. and Foxall, R. A., *A detailed study of the deformation of high purity niobium single crystals.* Phil. Mag. A, 1969. **20**: p. 719-751.

28.	Statham, C. D., Vesely, D., and Christian, J. W., *Slip in single crystals of niobium-molybdenum alloys deformed in compression.* Acta Metall., 1970. **18**: p. 1243-1252.

29.	Irwin, G. J., Guiu, F., and Pratt, P. L., *Influence of orientation on slip and strain hardening of molybdenum single crystals.* Phys. Status Solidi A, 1974. **22**: p. 685-698.

30. Nawaz, M. H. A. and Mordike, B. L., *Slip geometry of tantalum and tantalum alloys.* Phys. Status Solidi A, 1975. **32**: p. 449-458.

31. Nagakawa, J. and Meshii, M., *Deformation of niobium single crystals at temperatures between 77 and 4. 2 K.* Philos. Mag. A, 1981. **44**: p. 1165-1191.

32. Hollang, L., Hommel, M., and Seeger, A., *The flow stress of ultra-high-purity molybdenum single crystals.* Phys. Status Solidi A, 1997. **160**: p. 329-354.

33. Hollang, L. and Seeger, A., *The flow-stress asymmetry of ultra-pure molybdenum single crystals.* Mater. Trans. JIM, 2000. **41**: p. 141.

34. Hollang, L., Brunner, D., and Seeger, A., *Work hardening and flow stress of ultrapure molybdenum single crystals.* Mater. Sci. Eng., A, 2001. **319-321**: p. 233-236.

35. Lassila, D. H., Goldberg, A., and Becker, R., *The effect of grain boundaries on the athermal stress of tantalum and tantalum-tungsten alloys.* Metall. Mater. Trans. A, 2002. **33**: p. 3457-3464.

36. Kirchner, H. O. K., *Plastic deformation of the alkali metals sodium and potassium.* Acta Phys. Austriaca, 1978. **48**: p. 111-129.

37. Basinski, Z. S., Duesbery, M. S., and Murty, G. S., *The orientation and temperature dependence of plastic flow in potassium.* Acta Metall., 1981. **29**: p. 801-807.

38. Siedersleben, M. E. and Taylor, G., *Slip systems in bcc Li-Mg alloys.* Philos. Mag. A, 1989. **60**: p. 631-647.

39. Duesbery, M. S. and Basinski, Z. S., *The flow stress of potassium.* Acta Metall. Mater., 1993. **41**: p. 643-647.

40. Pichl, W. and Krystian, M., *The flow stress of high purity alkali metals.* Phys. Status Solidi A, 1997. **160**: p. 373-383.

41. Krystian, M. and Pichl, W., *Investigation of the slip geometry of high-purity potassium by in situ x-ray diffraction.* Mater. Sci. Eng., A, 2004. **387-389**: p. 115-120.

42. Christian, J. W. and Vitek, V., *Dislocations and stacking faults.* Rep. Prog. Phys., 1970. **33**: p. 307-411.

43. Kubin, L. P., *Reviews on the deformation behavior of materials.* Reviews on the deformation behavior of materials, 1982. **4**: p. 181-275.

44. Christian, J., *Some surprising features of the plastic deformation of body-centered cubic metals and alloys.* Metall. Mater. Trans. A, 1983. **14**: p. 1237-1256.

45. Taylor, G., *Thermally-activated deformation of bcc metals and alloys.* Prog. Mater Sci., 1992. **36**: p. 29-61.

46. Seeger, A., J Phys IV, 1995. **5**: p. 45-65.

47. Pichl, W., *Slip geometry and plastic anisotropy of body-centered cubic metals.* Phys. Status Solidi A, 2002. **189**: p. 5-25.

48. Bassani, J. L., *Plastic flow of crystals.* Advances in Applied Mechanics, 1994. **30**: p. 191-257.

49. Seeger, A., *The temperature and strain-rate dependence of the flow stress of body-centred cubic metals: A theory based on kink-kink interactions.* Z. Metallkde, 1981. **72**: p. 369-380.

50. Suzuki, H., in *Dislocation dynamics*, Rosenfield, A. R., Editor 1968, Macgraw-Hill. p. 679.

51. Foxall, R. A. and Statham, C. D., *Dislocation arrangements in deformed single crystals of niobium-molybdenum alloys and niobium-9% rhenium.* Acta Metall., 1970. **18**: p. 1147-1158.

52. Hirth, J. P. and Lothe, J., *Theory of dislocations.* 1982, New York: Wiley.

53. Duesbery, M. S., *Dislocations in solids.* 1989, Amsterdam: North-Holland.

54. Duesbery, M. S., *The dislocation core and plasticity.* Dislocations in Solids, 1989. **8**: p. 67-173.

55. Yip, S., *Handbook of materials modeling.* 2005, New York: Spinger.

56. Vitek, V. and Paidar, V., in *Dislocations in solids*, Hirth, J. P., Editor 2008, Elsevier: Amsterdam. p. 439.

57. Allen, N. P., Hopkings, B. E., and Mclennan, L. E., *The tensile properties of single crystals of high-purity iron at temperatures from 100 to -253 C.* Proc. R. Soc. Lond. A., 1956. **234**: p. 221-246.

58. Basinski, Z. S. and Christian, J., *The influence of temperature and strain rate on the flow stress of annealed and decarburized iron at subatmospheric temperatures.* Austr. J. Phys, 1960. **13**: p. 299-308.

59. Conrad, H. and Schoeck, G., *Cottrell locking and the flow stress in iron.* Acta Mater., 1960. **8**: p. 791-796.

60. Conte, R., Groh, P., and Escaig, B., *Étude de l'écoulement plastique de monocristaux filamentaires de fer entre 300 et 4°k par.* Phys. Status Solidi B., 1968. **28**: p. 475-488.

61. Conrad, H. and Frederik, S., *The effect of temperature and strain rate on the flow stress of iron.* Acta Metal., 1962. **10**: p. 1013-1020.

62. Stein, D. T., Low Jr., J. R., and Seybolt, A. U., *The mechanical properties of iron single crystals containing less than* $5 \times 10E\text{-}3$ *ppm carbon.* Acta Metall., 1963. **11**: p. 1253-1262.

63. Stein, D. T. and Low Jr., J. R., *Effects of orientation and carbon on the mechanical properties of iron single crystals.* Acta Metall., 1966. **14**: p. 1183-1194.

64. Arsenault, R. J., *Low-temperature creep of alpha iron.* Acta. Matall., 1964. **12**: p. 547-554.

65. Keh, A. S. and Nakada, Y., *Plasticity of iron single crystals.* Canad. J. Phys., 1967. **45**: p. 1101-1120.

66. Spitzig, W. A. and Keh, A. S., *Orientation dependence of the strain-rate sensitivity and thermally activated flow in iron single crystals.* Acta Metall., 1970. **18**: p. 1021-1033.

67. Spitzig, W. A., *Analysis of thermally-activated flow in iron single crystals.* Acta Metall., 1970. **18**: p. 1275-1284.

68. Aono, Y., Kuramoto, E., and Kitajima, K., *Reports of research institute for applied mechanics*, 1981, Kyushu University. p. 127-193.

69. Brunner, D. and Diehl, J., *The use of stress-relaxation measurements for investigations on the flow-stress of alpha-iron.* Phys. Status Solidi A, 1987. **104**: p. 145-155.

70. Brunner, D. and Diehl, J., *Extension of measurements of the tensile flow-stress of high-purity alpha-iron single-crystals to very low-temperatures.* Zeitschrift Fur Metallkunde, 1992. **83**: p. 828-834.

71. Seeger, A., in *Dislocations*, Veyssiere, P., Kubin, L., and Castaing, J., Editors. 1984, C.N.R.S.: Paris. p. 141.

72. Brunner, D. and Diehl, J., *Temperature and strain-rate dependence of the tensile flow-stress of high-purity alpha-iron below 250-k .2. Stress temperature regime-II and its transitions to regime-I and regime-III.* Phys. Status Solidi A, 1991. **125**: p. 203-216.

73. Havner, K. S., *Finite plastic deformation of crystalline solids*, in *Cambridge monographs on mechnics and applied mathematics*, 1992, Cambridge University Press: Cambridge.

74. Gough, H. J., *The behaviour of a single crystal of alpha-iron subjected to alternating torsional stresses.* Proc. R. Soc. Lond. A, 1928. **118**: p. 498-534.

75. Barrett, C. S., Ansel, G., and Mehl, R. F., *Slip, twinning and cleavage in iron and silicon ferrite.* Trans. Am. Soc. Met., 1937. **25**: p. 702-733.

76. Chen, N. K. and Maddin, R., *Slip planes and the energy of dislocations in a body centered cubic structure.* Acta Metall., 1954. **2**: p. 49-51.

77. Seeger, A., *Why anomalous slip in body-centred cubid metals?* Mat. Sci. Eng. A, 2001. **A319-321**: p. 254-260.

78. Seeger, A. and Wasserbaech, W., *Anomalous slip - a feature of high-purity body-centred cubic metals.* Phys. Stat. Sol. A, 2002. **189**: p. 27-50.

79.   Sigle, W., *High-resolution electron microscopy and molecular dynamics study of the (a/2)[111] screw dislocation in molybdenum.* Philos. Mag. A, 1999. **79**: p. 1009-1020.

80.   Mendis, B. G., et al., *Use of the nye tensor in analyzing hrem images of bcc screw dislocations.* Philos. Mag., 2006. **86**: p. 4607-4640.

81.   Vitek, V., Perrin, R. C., and Bowen, D. K., *The core structure of 1/2(111) screw dislocations in bcc crystals.* Phil. Mag., 1970. **21**: p. 1049-1073.

82.   Basinski, Z. S., Duesbery, M. S., and Taylor, R., *Influence of shear stress on screw dislocations in a model sodium lattice.* Can. J. Phys., 1971. **49**.

83.   Neumann, F., *Vorlesungen über die theorie der elasticität* 1885, Leipzig.

84.   Vitek, V., *Theory of the core structures of dislocations in body-centered-cubic metals.* Cryst Lattice Defects, 1974. **5**: p. 1-34.

85.   Duesbery, M. S. and Richardson, G. Y., *Dislocation core in crystalline materials.* Critical Reviews in Solid State and Materials Sciences, 1991. **17**: p. 1-46.

86.   Duesbery, M. S., Vitek, V., and Cserti, J., *Non-schmid plastic behaviour in bcc metals and alloys.* Understanding materials, a festschrift for Sir Peter hirsch, ed. Humphreys, C. J.2002, London: The Institute of Materials. 165-192.

87.   Moriarty, J. A., et al., *Atomistic simulations of dislocations and defects.* J. Comput.-Aided Mater. Des., 2002. **9**: p. 99-132.

88.   Duesbery, M. S. and Vitek, V., *Overview no. 128: Plastic anisotropy in bcc Transition metals.* Acta Mater., 1998. **46**: p. 1481-1492.

89.   Ito, K. and Vitek, V., *Atomistic study of non-schmid effects in the plastic yielding of bcc metals.* Philos. Mag. A, 2001. **81**: p. 1387-1407.

90.   Mendelev, M. I., et al., *Development of new interatomic potentials appropriate for crystalline and liquid iron.* Philos. Mag., 2003. **83**: p. 3977-3994.

91.   Domain, C. and Monnet, G., *Simulation of screw dislocation motion in iron by molecular dynamics simulations.* Phys. Rev. Lett., 2005. **95**: p. 215506.

92.   Chaussidon, J., Fivel, M., and Rodney, D., *The glide of screw dislocations in bcc Fe: Atomistic static and dynamic simulations.* Acta Mater., 2006. **54**: p. 3407-3416.

93.   Ismail-Beigi, S. and Arias, T. A., *Ab initio study of screw dislocations in Mo and Ta: A new picture of plasticity in bcc transition metals.* Phys. Rev. Lett., 2000. **84**: p. 1499-1502.

94. Woodward, C. and Rao, S. I., *Ab-initio simulation of isolated screw dislocations in bcc Mo and Ta.* Philos. Mag. A, 2001. **81**: p. 1305-1316.

95. Woodward, C. and Rao, S. I., *Flexible ab initio boundary conditions: Simulating isolated dislocations in bcc Mo and Ta.* Phys. Rev. Lett., 2002. **88**: p. 216402.

96. Frederiksen, S. L. and Jacobsen, K. W., *Density functional theory studies of screw dislocation core structures in bcc metals.* Philos. Mag., 2003. **83**: p. 365-375.

97. Segall, D. E., et al., *Ab initio and finite-temperature molecular dynamics studies of lattice resistance in tantalum.* Phys. Rev. B, 2003. **68**: p. 014104.

98. Mrovec, M., et al., *Bond-order potential for molybdenum: Application to dislocation behavior.* Phys. Rev. B, 2004. **69**: p. 094115.

99. Mrovec, M., et al., *Bond-order potential for simulations of extended defects in tungsten.* Physical Review B, 2007. **75**: p. 104119.

100. Mrovec, M., et al., *Magnetic bond-order potential for iron.* Phys. Rev. Lett., 2011. **106**: p. 246402.

101. Gröger, R., Bailey, A. G., and Vitek, V., *Multiscale modeling of plastic deformation of molybdenum and tungsten: I. Atomistic studies of the core structure and glide of 1/2<111> screw dislocations at 0K.* Acta Mater., 2008. **56**: p. 5401-5411.

102. Duesbery, M. S., *On non-glide stresses and their influence on the screw dislocation core in body-centred cubic metals. I. The peierls stress.* Proc. R. Soc. Lond. A, 1984. **392**: p. 145-173.

103. Hill, R., *Continuum micro-mechanics of elastoplastic polycrystals.* J. Mech. Phys. Solids, 1965. **13**: p. 89-101.

104. Rice, J. R., *Inelastic constitutive relations for solids: An internal-variable theory and its application to metal plasticity.* J. Mech. Phys. Solids, 1971. **19**: p. 433-455.

105. Hill, R. and Havner, K. S., *Perspectives in the mechanics of elastoplastic crystals.* J. Mech. Phys. Solids, 1982. **30**: p. 5-22.

106. Hill, R. and Rice, J. R., *Constitutive analysis of elastic-plastic crystals at arbitrary strain.* J. Mech. Phys. Solids, 1972. **20**: p. 401-413.

107. Qin, Q. and Bassani, J. L., *Non-associated plastic flow in single crystals.* J. Mech. Phys. Solids, 1992. **40**: p. 835-862.

108. Qin, Q. and Bassani, J. L., *Non-schmid yield behavior in single crystals.* J. Mech. Phys. Solids, 1992. **40**: p. 813-833.

109. Gröger, R., et al., *Multiscale modeling of plastic deformation of molybdenum and tungsten: II. Yield criterion for single crystals*

based on atomistic studies of glide of 1/2<111> screw dislocations.
Acta Mater., 2008. **56**: p. 5412-5425.

110. Nabarro, F. R. N., *Dislocations in a simple cubic lattice* Proc. Phys. Soc., 1947. **59**: p. 256-272.

111. Friedel, J., *Dislocations.* 1964, Oxford: Pergamon.

112. Nabarro, F. R. N., *Theory of crystal dislocations.* 1967, Oxford: Oxford University Press.

113. Peierls, R. E., *The size of a dislocation.* Proc. Phys. Soc., 1940. **52**: p. 34-37.

114. Seeger, A., *On the theory of the low-temperature internal friction peak observed in metals.* Phil. Mag., 1956. **1**: p. 651-662.

115. Seeger, A. and Schiller, P., *Bildung und diffusion von kinken als grundprozess der versetzungsbewegung bei der messung der inneren reibung.* Acta Metall., 1962. **10**: p. 348-357.

116. Dorn, J. E. and Rajnak, S., *Nucleation of kink pairs and the peierls mechanism of plastic deformation.* Trans. AIME, 1964. **230**: p. 1052-1064.

117. Guyot, P. and Dorn, J. E., *A critical review of the peierls mechanism.* Canad. J. Phys., 1967. **45**: p. 983-1016.

118. Brunner, D. and Diehl, J., *Strain-rate and temperature-dependence of the tensile flow-stress of high-purity alpha-iron above 250K (regime I) studied by means of stress-relaxation tests.* Phys. Status Solidi A, 1991. **124**: p. 155-170.

119. Brunner, D. and Diehl, J., *Temperature and strain-rate dependence of the tensile flow-stress of high-purity alpha-iron below 250K .1. Stress temperature regime III.* Phys. Status Solidi A, 1991. **124**: p. 455-464.

120. Kocks, U. F., Argon, A. S., and Ashby, M. F., *Thermodynamics and kinetics of slip.* Prog. Mater. Sci., 1975. **19**: p. 1-291.

121. Suzuki, T., Takeuchi, S., and Yoshinaga, H., *Dislocation dynamics and plasticity.* 1985, Berlin: Springer.

122. Caillard, D. and Martin, J., *Thermally activated mechanisms in crystal plasticity.* 2003, Oxford: Elsevier-Pergamon.

123. Christian, J., *The theory of transformations in metals and alloys.* 2002, Oxford: Elsevier-Pergamon.

124. Seeger, A. and Sestak, B., Z. Metallk, 1978. **69**: p. 195;355;425.

125. Duesbery, M. S., *On kinked screw dislocations in the bcc lattice--I. The structure and peierls stress of isolated kinks.* Acta Metall., 1983. **31**: p. 1747-1758.

126. Duesbery, M. S., *On kinked screw dislocations in the bcc lattice--II. Kink energies and double kinks.* Acta Metall., 1983. **31**: p. 1759-1770.

127. Henkelman, G., Johannesson, G., and Jonsson, H., *Methods for finding saddle points and minimum energy paths.* Progress on Theoretical Chemistry and Physics, 2000: p. 269-300.

128. Henkelman, G. and Jonsson, H., *Improved tangent estimate in the nudged elastic band method for finding minimum energy paths and saddle points.* J. Chem. Phys., 2000. **113**: p. 9978-9985.

129. Henkelman, G., Uberuaga, B. P., and Jonsson, H., *A climbing image nudged elastic band method for finding saddle points and minimum energy paths.* J. Chem. Phys., 2000. **113**: p. 9901-9904.

130. Jonsson, H., Mills, G., and Jacobsen, K. W., *Nudged elastic band method for finding minimum energy paths of transitions.* Classical and Quantum Dynamics in Condensed Phase Simulations, 1998: p. 385-404.

131. Ngan, A. H. W. and Wen, M., *Atomistic simulation of energetics of motion of screw dislocations in bcc Fe at finite temperatures.* Comput. Mater., 2002. **23**: p. 139-145.

132. Wen, M. and Ngan, A. H. W., *Atomistic simulation of kink-pairs of screw dislocations in body-centred cubic iron.* Acta Mater., 2000. **48**: p. 4255-4265.

133. Rodney, D., *Activation enthalpy for kink-pair nucleation on dislocations: Comparison between static and dynamic atomic-scale simulations.* Physical Review B, 2007. **76**: p. 144108.

134. Rodney, D. and Proville, L., *Stress-dependent peierls potential: Influence on kink-pair activation.* Physical Review B, 2009. **79**: p. 094108.

135. Marian, J., Cai, W., and Bulatov, V. V., *Dynamic transitions from smooth to rough to twinning in dislocation motion.* Nat. Mater., 2004. **3**: p. 158-163.

136. Voter, A. F., Montalenti, F., and Germann, T. C., *Extending the time scale in atomistic simulation of materials.* Annu. Rev. Mater. Sci., 2002. **32**: p. 321-346.

137. Edagawa, K., Suzuki, T., and Takeuchi, S., *Motion of a screw dislocation in a two-dimensional peierls potential.* Phys. Rev. B, 1997. **55**: p. 6180.

138. Gröger, R. and Vitek, V., *Multiscale modeling of plastic deformation of molybdenum and tungsten. III. Effects of temperature and plastic strain rate.* Acta Mater., 2008. **56**: p. 5426-5439.

139. Tang, M., Kubin, L. P., and Canova, G. R., *Dislocation mobility and the mechanical response of bcc single crystals: A mesoscopic approach.* Acta. Mater., 1998. **9**: p. 3221-3235.

140. Cottrell, A. H. and Pettifor, D. G., *Electron theory in alloy design.* 1992, London: Institute of Materials, Minerals and Mining.

141. Hafner, J., *Atomic-scale computational materials science.* Acta Mater., 2000. **48**: p. 71-92.

142. Woodward, C. *First-principles simulations of dislocation cores.* in *Intenational Conference on Fundamentals of Plastic Deformation.* 2004. La Colle sur Loup, FRANCE: Elsevier Science Sa.

143. Finnis, M. W. and Sinclair, J. E., *Simple empirical n-body potential for transition metals.* Philos. Mag. A, 1984. **50**: p. 45-55.

144. Daw, M. S. and Baskes, M. I., *Semiempirical, quantum mechanical calculation of hydrogen embrittlement in metals.* Phys. Rev. Lett., 1983. **50**: p. 1285-1288.

145. Daw, M. S. and Baskes, M. I., *Embedded-atom method: Derivation and application to impurities and other defects in metals.* Phys. Rev. B, 1984. **29**: p. 6443-6453.

146. Andersen, O. K., *Simple approach to the band-structure problem.* Solid State Comm, 1973. **13**: p. 133-136.

147. Slater, J. C. and Koster, G. F., *Simplified LCAO method for the periodic potential problem.* Phys. Rev., 1954. **94**: p. 1498-1524.

148. Horsfield, A. P., et al., *Bond-order potentials: Theory and implementation.* Physical Review B, 1996. **53**: p. 12694.

149. Aoki, M., et al., *Atom-based bond-order potentials for modelling mechanical properties of metals.* Prog. Mater Sci., 2007. **52**: p. 154-195.

150. Znam, S., et al., *Atomistic modelling of TiAl I. Bond-order potentials with environmental dependence.* Philos. Mag., 2003. **83**: p. 415-438.

151. Stoner, E. C., *Collective electron ferromagnetism.* Proc. Roy. Soc. London A, 1938. **165**: p. 372–414.

152. Stoner, E. C., *Collective electron ferromagnetism: II. Energy and specific heat.* Proc. Roy. Soc. London A, 1939. **169**: p. 339-371.

153. Liu, G., et al., *Magnetic properties of point defects in iron within the tight-binding-bond stoner model.* Physical Review B, 2005. **71**: p. 174115.

154. Goodwin, L. and et al., *Generating transferable tight-binding parameters: Application to silicon.* EPL (Europhysics Letters), 1989. **9**: p. 701-706.

155. Urban, A., et al., *Parameterization of tight-binding models from density functional theory calculations.* Phys. Rev. B, 2011. **84**: p. 155119.

156. Ackland, G., et al., *Development of an interatomic potential for phosphorus impurities in α-iron.* J. Phys.: Condens. Matter, 2004. **16**: p. 2629-2642.

157. Ventelon, L. and Willaime, F., *Core structure and peierls potential of screw dislocations in alpha-Fe from first principles: Cluster versus dipole approach.* J. Computer-Aided Mater. Des., 2007. **14**: p. 85-94.

158. Caillard, D., *Kinetics of dislocations in pure Fe. Part II. In situ straining experiments at low temperature.* Acta Mater., 2010. **58**: p. 3504-3515.

159. Eyring, H. and Polanyi, M., *Ueber einfache gasreaktionen.* Z. Phy. Chem., 1931. **B12**: p. 279-311.

160. Laidler, K. and King, C., *Development of transition-state theory.* J. Phys. Chem., 1983. **87**: p. 2238-2256.

161. Mansoori, A., *Principles of nanotechnology: Molecular-based study of condensed matter in small systems.* 2005, MA: World Scientific.

162. Quapp, W. and Heidrich, D., *Analysis of the concept of minimum enery path on the potential energy surface of chemically reacting systems.* Theo. Chem. Acc., 1984. **66**: p. 245-260.

163. Jonsson, H., Mills, G., and Jacobsen, K. W., *Classical and quantum dynamics in condensed phase systems.* ed. Berne, B. J., Cicotti, G., and Coker, D. F.1998: World Scientific.

164. Olsen, R., et al., *Comparison of methods for finding saddle points without knowledge of the final states.* J. Chem. Phys., 2004. **121**: p. 9776-9792.

165. Henkelman, G., Johannesson, G., and Jonsson, H., *Theoretical ethods in condensed phase chemistry,* in *Theoretical chemistry and physics*2002, Springer: Netherlands. p. 269-302.

166. Schlegel, H., *Exploring potential energy surfaces for chemical reactions: An overview of some practical methods.* J. Comp. Chem, 2003. **24**: p. 1514-1527.

167. Galvan, I. and Field, M., *Improving the efficiency of the NEB reaction path finding algorithm.* J. Comp. Chem, 2008. **29**: p. 139-143.

168. Trygubenko, S. and Wales, D., *A doubly nudged elastic band method for finding transition states.* J. Chem. Phys., 2004. **120**: p. 2082-2094.

169. Bitzek, E., et al., *Structural relaxation made simple.* Phys. Rev. Lett., 2006. **97**: p. 170201.

170. Vitek, V., *Structure of dislocation cores in metallic materials and its impact on their plastic behaviour.* Prog. Mater Sci., 1992. **36**: p. 1-27.

171. Yang, L. H., Soderlind, P., and Moriarty, J. A., *Atomistic simulation of pressure dependent screw dislocation properties in bcc tantalum.* Mat. Sci. Eng. A, 2001. **309-310**: p. 102-107.

172. Bassani, J. L., Ito, K., and Vitek, V., *Complex macroscopic plastic flow arising from non-planar dislocation core structures.* Mater. Sci. Eng., A, 2001. **319-321**: p. 97-101.

173. Vitek, V., Mrovec, M., and Bassani, J. L., *Influence of non-glide stresses on plastic flow: From atomistic to continuum modeling.* Mater. Sci. Eng., A, 2004. **365**: p. 31-37.

174. Vitek, V., et al., *Effects of non-glide stresses on the plastic flow of single and polycrystals of molybdenum.* Mater. Sci. Eng., A, 2004. **387-389**: p. 138-142.

175. Peach, M. O. and Koehler, J. S., *The forces exerted on dislocations and the stress fields produced by them.* Phys. Rev., 1950. **80**: p. 436-439.

176. Vitek, V., in *Dislocations and properties of real materials*, Lorretto, M., Editor 1985, The Institute of Metals: London. p. 30.

177. Vitek, V., in *Handbook of materials modeling part B: Models*, Yip, S., Editor 2005, Springer: New York. p. 2883.

178. Takeuchi, S. and Kuramoto, E., *Thermally activated motion of a screw dislocation in a model bcc crystal.* J. Phys. Soc. Japan, 1975. **38**: p. 480-487.

179. Koizumi, H., Kirchner, H. O. K., and Suzuki, T., *Kink pair nucleation and critical shear stress.* Acta Metall. Mater., 1993. **41**: p. 3483-3493.

180. Suzuki, T., Kamimura, Y., and Kirchner, H. O. K., *Plastic homology of bcc metals.* Philos. Mag. A, 1999. **79**: p. 1629-1642.

181. Suzuki, T., Koizumi, H., and Kirchner, H. O. K., *Plastic flow stress of bcc transition metals and the peierls potential.* Acta Metall. Mater., 1995. **43**: p. 2177-2187.

182. Takeuchi, S., *Core structure of a screw dislocation in the bcc lattice and its relation to slip behaviour of α-iron.* Phil. Mag. A, 1979. **39**: p. 661-671.

183. Terentyev, D. A. and Malerba, L., *Effect of Cr atoms on the formation of double kinks in screw dislocations in Fe and its correlation with solute hardening and softening in FeCr alloys.* Comput. Mater., 2008. **43**: p. 855-866.

184. Dagens, L., Rasolt, M., and Taylor, R., *Charge-densities and interionic potentials in simple metals - nonlinear effects .2.* Phys. Rev. B, 1975. **11**: p. 2726-2734.

185. Moriarty, J. A., *Analytic representation of multiion interatomic potentials in transition-metals.* Phys. Rev. B, 1990. **42**: p. 1609-1628.

186. Gröger, R. and Vitek, V., *Breakdown of the schmid law in bcc molybdenum related to the effect of shear stress perpendicular to the slip direction.* Mater. Sci. Forum, 2005. **482**: p. 123-126.

187. Gröger, R. and Vitek, V., *Explanation of the discrepancy between the measured and atomistically calculated yield stresses in body-centred cubic metals.* Philos. Mag. Lett., 2007. **87**: p. 113 - 120.

188. Spitzig, W. A. and Keh, A. S., *Orientation and temperature dependence of slip in iron single crystals.* Metall. Trans., 1970. **1**: p. 2751-2757.

189. Nine, H. D., J. Appl. Phys., 1973. **44**: p. 4875-4881.

190. Etemad, B. and Guiu, F., *Flow stress asymmetry in cyclic deformation of molybdenum crystals.* Scr. Metall., 1974. **8**: p. 931-935.

191. Anglada, M., et al., *Stress asymmetry and shape changes in cyclically deformed Mo single crystals.* Scr. Metall., 1980. **14**: p. 1319-1322.

192. Seeger, A. and Hollang, L., *The flow-stress asymmetry of ultra-pure molybdenum single crystals.* Mater. Trans. JIM, 2000. **41**: p. 141-151.

193. Anglada, M. and Guiu, F. in *5th Int. Conf. on Strength of Metals and Alloys.* 1979. Aachen: Pergamon Press.

194. Anglada, M. and Guiu, F., *Cyclic deformation of Nb single crystals I. Influence of temperature and strain rate on cyclic hardening, shape changes and stress asymmetry.* Philos. Mag. A, 1981. **44**: p. 499 - 522.

195. Anglada, M. and Guiu, F., *Cyclic deformation of Nb single crystals II. Influence of orientation on cyclic hardening, shape changes and stress asymmetry.* Philos. Mag. A, 1981. **44**: p. 523 - 541.

196. Chang, L. N., Taylor, G., and Christian, J. W., *A stress asymmetry in niobium single crystals deformed at 77 K.* Scr. Metall., 1982. **16**: p. 95-98.

197. Chang, L. N., Taylor, G., and Christian, J. W., *Stress asymmetries in the deformation behaviour of niobium single crystals.* Acta Metall., 1983. **31**: p. 37-42.

198. Koss, D. A. *Mechanical properties of bcc metals.* in *TMS-AIME.* 1982. Warrendale, PA.

199. Mott, N. F., *Creep in metal crystals at very low temperatures.* Phil. Mag., 1956. **1**: p. 568-572.

200. Weertman, J., *Dislocation model of low - temperature creep.* J. Appl. Phys., 1958. **29**: p. 1685-1689.

201. Gilman, J., *Escape of dislocations from bound states by tunneling.* J. Appl. Phys., 1968. **39**: p. 6086-6090.

202. Petukhov, B. V. and Pokrovskii, V. L., *Quantum and classical motion of dislocations in a Peierls potential relief.* Sov. Phys. JETP, 1972. **36**.

203. Leibfried, G., *Dislocation and mechanical properties of crystals.* 1957, New York: John Wiley.

204. Alefeld, G., *Rate theory in solids at low temperatures.* Phys. Rev. Lett., 1964. **12**: p. 372-375.

205. Suzuki, H., *Fundamental aspects of dislocation theory.* 1970: U. S. National Bureau Standards.

206. Nastik, V. D., et al., Phys. Stat. Sol. B 1972. **34**: p. 99.

207. Suzuki, T. and Ishii, T., *Physics of strength and plasticity.* 1969, Cambridge: MIT Press.

208. Suenaga, M. and Galligan, J. M., in *Physical acoustics*, Mason, W. P., Editor 1972, Academic Press: NewYork. p. 1.

209. Brunner, D. and Diehl, J., *The effect of atomic lattice defects on the softening phenomena of high-purity alpha-iron.* Phys. Status Solidi A, 1997. **160**: p. 355-372.

210. Kuramoto, E., Aono, Y., and Kitajima, K., *Thermally activated slip deformation of high purity iron single crystals between 4.2 K and 300 K.* Scr. Metall., 1979. **13**: p. 1039-1042.

Schriftenreihe
des Instituts für Angewandte Materialien

ISSN 2192-9963

Die Bände sind unter www.ksp.kit.edu als PDF frei verfügbar oder
als Druckausgabe bestellbar.

Band 1   Prachai Norajitra
          Divertor Development for a Future Fusion Power Plant. 2011
          ISBN 978-3-86644-738-7

Band 2   Jürgen Prokop
          **Entwicklung von Spritzgießsonderverfahren zur Herstellung von
          Mikrobauteilen durch galvanische Replikation.** 2011
          ISBN 978-3-86644-755-4

Band 3   Theo Fett
          **New contributions to R-curves and bridging stresses – Applications
          of weight functions.** 2012
          ISBN 978-3-86644-836-0

Band 4   Jérôme Acker
          **Einfluss des Alkali/Niob-Verhältnisses und der Kupferdotierung auf
          das Sinterverhalten, die Strukturbildung und die Mikrostruktur von
          bleifreier Piezokeramik $(K_{0,5}Na_{0,5})NbO_3$.** 2012
          ISBN 978-3-86644-867-4

Band 5   Holger Schwaab
          **Nichtlineare Modellierung von Ferroelektrika unter Berücksich-
          tigung der elektrischen Leitfähigkeit.** 2012
          ISBN 978-3-86644-869-8

Band 6   Christian Dethloff
          **Modeling of Helium Bubble Nucleation and Growth in Neutron
          Irradiated RAFM Steels.** 2012
          ISBN 978-3-86644-901-5

Band 7   Jens Reiser
          **Duktilisierung von Wolfram. Synthese, Analyse und Charakterisierung
          von Wolframlaminaten aus Wolframfolie.** 2012
          ISBN 978-3-86644-902-2

Band 8   Andreas Sedlmayr
          Experimental Investigations of Deformation Pathways in Nanowires.
          2012
          ISBN 978-3-86644-905-3

Band 9   Matthias Friedrich Funk
             **Microstructural stability of nanostructured fcc metals during cyclic
             deformation and fatigue.** 2012
             ISBN 978-3-86644-918-3

Band 10   Maximilian Schwenk
             **Entwicklung und Validierung eines numerischen Simulationsmodells
             zur Beschreibung der induktiven Ein- und Zweifrequenzrandschicht-
             härtung am Beispiel von vergütetem 42CrMo4.** 2012
             ISBN 978-3-86644-929-9

Band 11   Matthias Merzkirch
             **Verformungs- und Schädigungsverhalten der verbundstranggepressten,
             federstahldrahtverstärkten Aluminiumlegierung EN AW-6082.** 2012
             ISBN 978-3-86644-933-6

Band 12   Thilo Hammers
             **Wärmebehandlung und Recken von verbundstranggepressten
             Luftfahrtprofilen.** 2013
             ISBN 978-3-86644-947-3

Band 13   Jochen Lohmiller
             **Investigation of deformation mechanisms in nanocrystalline
             metals and alloys by in situ synchrotron X-ray diffraction.** 2013
             ISBN 978-3-86644-962-6

Band 14   Simone Schreijäg
             **Microstructure and Mechanical Behavior of Deep Drawing DC04 Steel at
             Different Length Scales.** 2013
             ISBN 978-3-86644-967-1

Band 15   Zhiming Chen
             Modelling the plastic deformation of iron. 2013
             ISBN 978-3-86644-968-8